The Garden in the Machine

The Garden
in the Machine

THE EMERGING SCIENCE
OF ARTIFICIAL LIFE

Claus Emmeche

Translated by Steven Sampson

PRINCETON UNIVERSITY PRESS

PRINCETON, NEW JERSEY

Copyright © 1994 by Princeton University Press
Published by Princeton University Press, 41 William Street,
Princeton, New Jersey 08540
In the United Kingdom: Princeton University Press,
Chichester, West Sussex

First published in Danish by Munksgaard as
Det Levende Spil: Biologisk Form og Kunstigt Liv.
© 1991 Claus Emmeche.

All Rights Reserved

Library of Congress Cataloging-in-Publication Data

Emmeche, Claus, 1956–
[Levende Spil. English]
The garden in the machine: the emerging science of artificial
life / Claus Emmeche; translated by Steven Sampson.
p. cm.
Translation of: Det Levende Spil.
Includes bibliographical references and index.
1. Biological systems—Computer simulation.   2. Biology—
Philosophy.   3. Life.   I. Title.
QH324.2.E4613   1994   577—dc20   93-39101

ISBN 0-691-03330-7

This book has been composed in Adobe Palatino

Princeton University Press books are printed
on acid-free paper and meet the guidelines for
permanence and durability of the Committee
on Production Guidelines for Book Longevity
of the Council on Library Resources

Printed in the United States of America

3 5 7 9 10 8 6 4 2

CONTENTS

PREFACE

"And the Nightingale sang so gloriously that the
tears came into the Emperor's eyes"
—Hans Christian Andersen, "The Nightingale"

THE DANISH physicist Niels Bohr has been called "the last
vitalist." Bohr believed that the "vitality" of organisms—
that they are living beings and not simply inert lumps of
matter—could not be conceptualized strictly by natural science alone. Bohr spoke of complementarity between the
physical description of life and the idea of the living as
something that was purposeful, undivided, and integrated,
something that could not be taken apart without destroying
the very quality of life.

This book deals with the use of computers as a new perspective on the problem of the vitality of life. One response
to this problem is "artificial life." It is a perspective that
acknowledges that life possesses certain specific "vital" fundamentals, but that denies their mystical character. The artificial-life orientation emphasizes that it ought to be possible to imitate or remake these fundamentals artificially, in
a computer, for instance. In order to understand this response, we must first understand the basic question of what
life is.

Today we know inordinately more about what happens in
a living cell, its physiology, genetics, and evolution, than
Niels Bohr did in 1932 when, in a lecture called "Light and
Life," he applied physics's notion of complementarity to biology. Yet our biological knowledge still has a fragmentary
character. Science seeks coherent insights. Besides the principles from evolutionary theory, which Darwin and his successors formulated, there are no natural laws in biology,

nor are there comprehensive mathematical theories that correspond to physics's quantum mechanics or relativity theory.

Modern biochemistry and molecular biology are often considered to be the ultimate defeat of vitalism. Vitalism is here considered a quasireligious belief that living organs contain a unique vital principle, a mystical life-force or something similar, that cannot be explained within the framework of natural science. As the belief in metaphysical principles or supernatural forces in living organisms, vitalism did not have an adherent in Niels Bohr. Bohr was content to say that a physical description has limited validity when it concerns understanding life or consciousness.

Molecular biology has not fundamentally altered this insight. Traditional biology has been analytical in its methods and has focused on the material basis for life. We take a living thing and dissect it in order to see what it consists of and how its parts fit together. We seek to derive the underlying principles behind their organization. Then one dissects *these* parts and examines what *they* consist of, and then their components, and so on. The method is adequate up to a certain point, for suddenly we are not describing anything living. The "life" being described suddenly slips through our fingers. Hence, the analytical description must be complemented by a synthetic one that leaves room for the more holistic features that characterize all living beings.

Today researchers who use molecular biology to transfer genes from one species to another are in fact making artificial life a reality. Think of the gene-spliced sugar beets, the supermice, or the microorganisms that are genetically designed for specific purposes in production. The public debate about artificial life-forms gives the impression that current biotechnology can design an organism totally from scratch. In fact, this is a fantastically misleading oversimplification of the situation. It is not a matter of creating life from scratch (de novo), atom by atom, molecule by molecule. Rather, we are genetically altering already-living organisms. Therefore, *artificially modified life* is a more precise designation for these gene-spliced organisms. Were we to start totally from scratch, we could not exclude the possibility that this experiment should repeat the entire, long, slow evolutionary process that has taken place on earth, from the time when the world's seas, before becoming complex ecosystems, were but a large soup of organic molecules. Were we to abbreviate the evolutionary game, it would demand a knowledge of the organization of a living organism that we do not yet have.

In principle, it is perhaps possible to make artificial life by chemical means, in a test tube (in vitro). Vitalism is dead yet again. The reality, however, is that we are very far from being able to create anything that at all exhibits the complexity of a living cell.

This does not mean that vitalism will return. Yet it certainly leads us to be cautious regarding the question of natural versus artificial life. Vitalism is certainly dead, but molecular biology has nearly died in its moment of triumph; despite the acquisition of a considerable corpus of detailed knowledge, it has produced only a limited number of genuine breakthroughs, especially concerning the theoretically important and unresolved questions in biology: the general questions about form, development, self-organization, and other media as possible bearers of life.

There exists, however, a radically different path towards artificial life. This effort can be compared with a mathematical description of Nature, but in an entirely new form. It is an old notion that nature can be understood mathematically—Newtonian physics was a grand attempt toward a mathematical re-creation (on paper) of the laws that govern the universe. Can the biological universe be understood in a similar way?

Today the computer is used in biology to make mathematical models of biological phenomena: one example is the fishery biologist's models of the North Sea ecosystem, a model that seeks to describe the growth in the different fish populations. These kinds of models, however, are only a fragment of the myriad of possibilities for creative scientific applications of the computer for understanding fundamental processes in our universe. It is simply not only possible to make mathematical *models* of biological systems. There is today an intellectual movement going on—with roots in biology, physics, computer science, and mathematics—that has brought researchers together in the attempt to *synthesize* life; that is, to realize the creation of life processes through the medium of the computer.

In the view of this intellectual movement, life is not a question of the different materials we consist of but of the *organization* of the elements in time and space, of the interaction of relations and processes of which these elements are a part. The material elements could be replaced by other types—they could, for example, be the small chips of silicon in the computer (in silico).

This book deals with the following radical notion: that life can be calculated because life in itself realizes general forms of movement, forms of processing, that are computational in nature. If life is a machine, the machine itself can become living. The computer can be the path to life.

In following my argument, the reader should not feel overcome by details that may at first seem difficult to grasp. I have attempted in this book to be as user-friendly as possible. Those who would like to pursue these ideas further can find some help in the accompanying endnotes.[1]

The main idea is easy to comprehend. This book is more a story about a scientific quest (a new research program) than about a final set of results. Whether the effort to create artificial life will succeed is impossible to say because it depends especially on what we accept as a fully valid example of authentic life.

If the research program for artificial life becomes successful, it will have enormous consequences for the way we view life generally. I have attempted to be as fair as possible to this potential new paradigm. Yet I will also describe the puzzlement and skepticism that efforts to create artificial life have aroused among ordinary people and among biologists themselves. Our immediate and essential intuition of the living is that life is earthly, bound up with the need for water and the circulation of carbon compounds.

In the midst of our enthusiasm for these new developments, it is worth recalling Niels Bohr's admonition. The idea of artificial life raises critical questions and creates general puzzlement: is there not a fundamental difference between a forest snail and a computer? Is life not something that is slimy and grubby? And can an artificial nightingale ever replace a real one?

ACKNOWLEDGMENTS

MANY PEOPLE helped me during the preparation of the original edition of this book, published in Danish as "Det Levende Spil" in 1991. For their help and inspiration, I wish to thank Niels Engelsted, Jacob Erle, Hans Siggaard Jensen, Mogens Kilstrup, Carsten Knudsen, Chris Langton, Michael May, Lone Malmborg, Erik Mosekilde, Ib Ravn, Robert Rosen, Jakob Skipper, Eliott Sober, Ib Ulbæk, and Jakob Zeuthen Dalgaard. For kind support and stimulating discussions during my subsequent work on artificial life, theoretical biology, and philosophy, thanks to Nils A. Baas, Niels Ole Bernsen, Jesper Hoffmeyer, Simo Køppe, Benny Lautrup, and Frederik Stjernfelt.

I also wish to thank my translator, Steven Sampson, for turning the Danish manuscript into readable English. Last but not least, thanks to my beloved wife Birte Olsen for diverting my attention in the most beautiful and creative way during the whole process.

· · ·

Fig. 2.1 reproduced from Wonderful Life: The Burgess Shale and the Nature of History by Stephen Jay Gould, illustrated by Marianne Collins. Reprinted by permission of W. W. Norton & Company, Inc. Copyright c 1989 by Stephen Jay Gould. Figs. 2.2, 3.2, 3.7, 4.1, 4.2, 6.2, 7.1, 7.2 reproduced courtesy of Jesper Tom-Peterson. Fig. 3.1 reproduced from Alfred Chapuis and Edmund Droz 1958: Automata: A historical and technological study (London: B. T. Batsford), figs. 286–87. Reprinted with permission of Editions du Griffon. Figs. 3.3a, 3.3b reproduced from Christopher Langton 1984: "Self-reproduction in cellular automata," Physica D 10:135–44. Reprinted with permission of Elsevier Science Publishers. Figs. 3.3c, 3.3d reproduced from Christopher Langton, ed., Artificial Life II (figs. 2(e) and 2(g), p. 824), c 1992 by

Addison-Wesley Publishing Company, Inc. Reprinted by permission. Fig. 4.3 reproduced from A. Lindenmeyer and P. Prusinkiewicz 1990: *The algorithmic beauty of plants* (New York: Springer-Verlag), p. 25. Reprinted with permission of Springer-Verlag. Fig. 4.4 reproduced from V. Pauchet and S. Dupres 1976: *Pocket atlas of anatomy* (Oxford: Oxford University Press), pl. 116. Reprinted by permission Oxford University Press. Fig. 4.5 reproduced from Stephen Wolfram 1984: "Universality and complexity in cellular automata," *Physica D* 10:1–35. Reprinted with permission of Elsevier Science Publishers and Stephen Wolfram. Fig. 4.6 reproduced from C. H. Waddington 1957: *The Strategy of Genes* (London: George Allen and Unwin). Reprinted with permission of HarperCollins Publishers, Ltd. Fig. 4.7 reproduced from Christopher Langton, ed., *Artificial Life* (fig. 8, p. 32), c 1989 by Addison-Wesley Publishing Company, Inc. Reprinted by permission. Fig. 5.2 reproduced from Christopher Langton 1990: "Computation at the edge of chaos," *Physica D* 42:12–37. Reprinted with permission of Elsevier Science Publishers.

The Garden in the Machine

Chapter One

THE GAME OF LIFE

"In fact, it's quite easy to create life."

The little man with the dark eyes smiles. He has just summarized his lecture. A weak buzz from his computer confirms his conclusion. We can create life—and not only by reproducing our own flesh and blood; rather, we can create completely new life. The man is Thomas Ray from the University of Delaware. He does not have any supernatural or religious capacity. As a trained ecologist, Ray knows his biology and has a rather unsentimental, down-to-earth relation to living things. The kind of life Ray is talking about creating is biological life, but of a very special kind.

We are at the Technical University of Denmark, just outside Copenhagen, where Ray is visiting a research group that studies chaos theory and artificial life. A few years ago, Ray had renounced the study of tropical ecosystems, which had been his specialty. Ray was not bored, nor had he cynically realized that his object of study was soon going to disappear and that he had to find a new niche in the academic world. No, no matter how much Ray loved his rain forests, there was something else that had attracted his attention. He had gotten a remarkable idea: that with the help of computers he could piece together fragments of computer programs (i.e., instructions) and turn them into artificial organisms that did not just resemble life, but that theoretically speaking were just as alive as real animals and plants.

The idea seems a bit insane, or rather grandiose. Today, however, Ray can conclude that it is in fact "quite easy" to reproduce life. Just think of the computer viruses that plague our computer systems. The viruses are often created by hackers, usually teenagers with an intimate knowledge of their own and others' computers. They hack themselves into company or government computer networks and then release their homemade computer programs or viruses into these giant systems. These virulent pieces of computer programming reproduce themselves and then spread to all the computers in the network. Here they often make complete chaos of the most sophisticated systems and erase important data.[2] Viruses are, in this sense, a form of life.

Ray was not the first to arrive at the idea of computer life. In 1989, when he began to work on creating organisms, scientists from many different disciplines were already testing the idea of creating life in artificial universes. A common thread began to emerge between them. This new life would be more fun, and it would be a more creative life than the destructive computer viruses that had wreaked havoc on several scientific centers. The project itself was to be interdisciplinary and scientifically ambitious, in the same way as in the 1950s when logicians, computer scientists, and psychologists came together in the attempt to construct "artificial intelligence."

A modest man, Ray entitled his project "An Approach to the Synthesis of Life," implying that there exist other possible methods as well. Life is a process, a complex, rhythmic pattern of matter and energy. What is important is not what kind of matter or what kind of energy we find, but rather the pattern, the process, the *form*. The computer's powers of information processing can imitate other forms and processes so well that the result is not simply an imitation or a theoretical image of life. The pulsating patterns on the computer screen are themselves new examples of how the process of life itself can take form, says Ray. Such self-developing patterns of processes *are* life.

There are an infinite number of processes in nature and in society where minute changes have profound, wide-ranging effects. If one removes the lowest can in a supermarket's tower of canned tuna fish, a large number of events occur quickly, one after the other; not only do the cans fall, but the tower itself vanishes. Physically speaking, it is easy enough to understand that a tower of canned tuna, which represents an organized system in an unstable balance, may quickly *evolve* into a more stable state: a pile of cans on the floor. This happens when the small change that takes place at the bottom of the tower ruptures the fragile symmetry and thereby converts the cans' potential energy into movement, heat, and a considerable amount of disorder. The cans come alive nearly all by themselves. But the cans' development in a supermarket does not have much to do with life in a biological sense. When we say that life is *self-organizing*, we refer to that which typifies a living organism (and the species' evolution), the organization that is built up and maintained rather than destroyed. Order does not disappear like the tower of cans in the supermarket; something is broken down along the way, to be sure, but order is continually recreated.

Ray discovered that he did not need to do much himself before all kinds of things began to show up on his computer. The events on the screen were more reminiscent of life's self-organization than of the spontaneous fall of a tower of cans. When Ray designed just one simple stem-organism, in the form of a program that he set off in his computer, it evolved into a veritable zoo of various species of descendants. Some were parasites (programs that utilize the computational resources of other programs). Others were hyperparasites (which exploited the parasites); and there also emerged more social organisms. All these organisms evolved at the cost of the machine's memory, but without Ray himself having to control their direction or their consumption of memory. Ray was startled to see the various new life-forms that appeared on the screen after he had released his homemade ancestral organism. I will later go into more detail about similar types of computer life.

That computers only do what they have been programmed to do has become a truism with modifications. All Ray did was to create a little "universe" with his machine, give it a few rules to work with, and then allow the system to evolve by itself. Just like Isaac Newton, who imagined that God, after having created the universe, the elements, and the eternal laws of motion, sat back and rested, without any further participation. Newton, however, did not regard the universe as being completely stable, so his God had to intervene from time to time in order to restore balance. Ray does not need to. Ray's microuniverse organizes itself, to a certain degree. Here it is not stability but evolution that really counts.

THE CASE OF THE FLICKERING GAME

In order to provide an initial, tangible notion of how artificial life behaves in a computer, it is easiest to examine a game called "Life," invented by the mathematician John Horton Conway in 1970. "Life" is so simple that it can be played on a chess board with a few pieces, pawns for instance. It is easier than chess, and the kinds of pieces used are in fact unimportant.

Each square on the board, regardless of color, is a "cell." A cell can have one of two states: "on" (when there is a piece on the square), or "off." Each cell has a total of eight neighboring cells (surrounding squares). The game is a solitary one; there is no opponent. It begins as follows:

Choose a beginning situation. For example, allow three cells in a row to be on and the remainder off. This is the beginning position. The game now proceeds step by step such that for each individual step, one calculates its next state (next generation) using two simple rules:

1. A cell is turned on if three of its neighbors are turned on
2. A cell remains on if two or three of its neighbors are also on; otherwise it is turned off.[3]

Conway formulated the rules slightly more organically and with more words:

1. Survival: each piece with two or three neighbors survives in the next generation
2. Death: each piece with four or more neighbors dies from

overpopulation, and each piece with one or no neighbors dies of isolation

3. Birth: each empty cell with precisely three inhabited neighbor cells is a birth cell, where a new piece is born in the next generation.

The choice of words, however, is not important as long as the logic of the game remains the same.

For each cell, one counts the number of neighbors and uses the rules in order to determine its state in the next generation. When this has been done for all cells and the pieces are placed correctly, one has in reality "updated the cellular automaton" (for a moment, we can think of the game as a mechanical device, an automaton). In other words, one has computed the state of the cells on the entire playing board for the next time-step. One can thereafter continue to calculate the next generation, and the next, and so on until one decides to stop.

All cells in a generation change their state (i.e., are born, survive, or die) simultaneously. This means that one must not remove the pieces in a field that neighbors a cell if the states of the next generation have not yet been calculated. Conway recommended the following technique:

1. Start with a pattern of black pieces
2. Identify those who must die and lay a black piece on top of them
3. Find the birth cells and lay a white piece in each
4. Checking and double-checking, remove the dead pieces and replace the newborn pieces with black ones.

One soon discovers that depending on the initial state chosen (i.e., how many cells are on and how dispersed they are from each other) quite different things begin to happen. Sometimes the game dies out rapidly, and all the cells are turned off. Groups of cells can be frozen solid in a crystalline state, a stiff pattern that cannot be changed (unless "pushed" by new patterns). However, there may also appear "wave" sequences that are impossible to maintain

within the boundaries of the chess board because the patterns grow; in this case more and more neighboring cells
are born, and it takes increasingly more time to calculate
the state of the entire next generation, the number of pieces
is exhausted, and so on.

Conway quickly encountered problems because instead
of a chess board he tried the game out in his home, on the
checked tiles in the large entrance hallway, and he used
dishes to mark the different cells. He quickly became tired
of this when he literally began to step on the pieces. Luck-
ily, the game can be played more effectively on a computer,
and the computer can also calculate the generation change
much more rapidly as well. Here the game really begins to
resemble something living: small vibrating patterns go back
and forth in wave-like rhythms, and the new structures
emerge from simple seedlings. One of these new structures
is a "glider," which crawls its way diagonally across the
chess board until it crashes into other structures and allows
itself to be swallowed up. A glider looks like this:

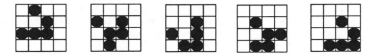

FIGURE 1.1. Microscopic section of John Conway's game "Life," in five
consecutive generations, where a "glider" moves diagonally down over
the network of cells. Along the way its form changes periodically as
shown. White means "off" (empty field), black means "on" (a piece on the
field). For each generation, a cell's state is determined according to the
rules cited above. Only sixteen cells are shown but the edge cells also have
eight neighbors, which cannot all be seen, and which are considered off.
The active cell in the highest row in the first generation (to the left) has
only one neighbor on; therefore it is off in the second generation, etc.

Conway and his colleagues gradually assembled a virtual
zoo of forms that could persist, send signals (like a "glider
gun"), keep blinking, or crawl around (more on this in
chapter 3). Copies of this simple program soon began to cir-
culate among computer enthusiasts, who themselves could
explore this world and discover within it new life-forms.

Conway's game is simple, but the structures that it generates can be quite complex. For followers of artificial life like Thomas Ray, the game of Life is a fully valid example of real life because it embodies the same principles of computation that are achieved in the more advanced forms of artificial life, including his own model.

SELF-ORGANIZATION:
DARWINISM'S UNRESOLVED QUESTION

If life in a computer really *is* life and it organizes itself, it can perhaps fulfill an old dream of biology. Some of the researchers who in the infant days of biology observed the growth of plants and the fetal development of animals dreamed of one day being able to explain the riddle of the generation of form and regulation of growth. They dreamed of a "Newton of the grassblade" who could do for biology what Sir Isaac Newton had done for physics.[4] Next to nothing was understood about the mechanics of life and metabolism; there were no formulas for a simple blade of grass. The ingenious functionality of organisms and their ability to live in ecologically marginal areas and adapt to harsh conditions could be explained only by invoking some kind of divine design. The sublime intricacy of life-forms seemed to be sufficient proof of the Creator's magnificence. Behold His masterful work!

After Darwin, in the middle of the last century, created his theory of evolution based on natural selection and finally published it, it was not long before Ernst Haeckel, one of his supporters, declared Darwin to be the Newton of biology. According to Haeckel, Darwin had achieved the fundamental physical explanation for the multiplicity of life. The wealth of life-forms had evolved via a long, gradual selection process, carried forward via the individual organism's struggle for survival. The divine design explanation was now unnecessary.

Even though several decades went by before Darwin's theory of natural selection was accepted, it later became established as a fundamental biological law, at least from the neo-Darwinian point of view. In principle, the theory is still valid today as the natural explanation. The problem, however, is that we have realized that it is only half the explanation.

Darwin's evolutionary theory can be characterized as a theory about survival and gradual genetic change (toward increased fitness) among types of organisms. It is a theory about organisms that are *already* partially well adapted, well-constructed, and highly organized biophysical systems. What Darwinian evolutionary theory fails to explain is how animals and plants were created originally. Natural selection did not just leave the gigantic riddle about the origin of life unsolved; Darwin's theory also offered no insight about the creation of the individual multicellular organism in each new generation. Darwin knew nothing about self-organization. His theories were decisive for the subsequent development of biology, but the lack of understanding about the dynamics of the creation of a living structure left theoretical biology in a quandary. This became a major stumbling block for twentieth-century research. It is only within the last two decades that we have really begun to discover some of the principles for the creation of form, the creation of the complex from the simple, that must supplement Darwinian theory.

There are many paths to this new understanding. One of them is embryology itself, which took on the task of minutely sketching out the phases in the development of the vertebrate embryo. Another path is that of molecular biology, which studies the mutations that disturb the normal developmental sequences. The sequence that a fetus follows has been called the "epigenetic landscape," i.e., the total possible paths of cell development, growth, and anatomical creation of form. New insights have also come from physical theories of self-organization, such as the creation of ordered patterns in matter driven by energy flowing through the system (such as, e.g., hexagonal convection cells in a liquid heated from below to produce coherent motion of ensembles of molecules, the so-called "Bernard instability" phenomenon that organizes the system spatially). Finally, there is the contribution of computer science, both as a tool to simulate some of the complex morphogenetic processes and as a source of various metaphors for programmed development and coding. Ray and his colleagues, who claim to have created artificial life, are participants in a much larger process of intellectual history, a process with profound consequences for our way of viewing biology and life.

BETWEEN MYSTICISM AND BANALITY: FRACTAL KNOWLEDGE

The news that we can create artificial life sounds nearly magical. People program the material, and miraculously viruses, worms, and ants appear. We put form into the formless, pump sense into the senseless, let dust form into a system—do we now control the miracle of life? The idea is indeed dizzying, but the dizziness is due more to the challenge to our conventional ideas than to artificial life itself. Is it the conceptual gap between living and dead matter (and between controlled behavior and spontaneity) that is artificial, that gives us goose bumps when we suddenly look into a new space full of living forms that begin to fill the gap?

The creature on the computer screen, however, develops from a quite simple point of departure. There is no reason to mystify either artificiality or system-imposed spontaneity. What is mystical is not life presented as a biological phenomenon, nor life presented as a computer-based imitation. The mystical, rather, is the life we ourselves live—our own fascinations, aspirations, and conceptions, as well as the myriad of feelings that are aroused in us when we encounter the unknown, or when we find ourselves confronting the collapse of outdated concepts.

For years biochemists, physiologists, and geneticists have investigated the processes that are linked to the wet insides of living cells. Since its advent we have used the computer as an aid for the storage and manipulation of data. Yet until very recently, the idea that real life could be synthesized, let alone using the computer as a substratum, was a totally foreign notion. Whereas we had previously viewed organisms as machines, we were now not only faced with the strange idea that synthetic life could in principle be created, but that it could be created from the computer's silicon chips, entirely without carbon compounds, the basic elements found in all life that we know of on earth.

The mechanistic tradition persists in small alcoves of biology and periodically appears in certain researchers' extemporaneous philosophizing. The mechanistic biologist believes that an organism is in reality nothing more than a collection of atoms, a simple machine made of organic molecules. Mainstream biology, however, asserts that physics, despite the undeniable physical and material properties of organisms, is nevertheless inadequately equipped to explain them. An organism's structure and function are phenomena that exist in their own right, and must be described with biology's own conceptual apparatus. (This rather mundane biological idea has been given the imposing name "organicism.")[5]

Nevertheless, biologists have often employed a range of metaphors to describe the real nature of organisms, and the metaphors have typically been borrowed from the technology that happened to be most fashionable at the moment. An ant, for example, can be viewed as a mechanical piece of clockwork, with precise, finely tuned parts, each with its distinct function. From a subsequent perspective, the ant can be viewed as a piece of energy technology: a thermodynamic design that—in analogy to a steam engine—consumes chemically bound energy by combustion and performs work while developing heat. Today we might view the ant as a little computer with associated sensory and motor organs: it processes a mass of information about the external world and reacts by feeding back various responses.

Regardless of how one views living beings—in terms of mechanics, thermodynamics, as information-processing machines, as entirely unique types of systems in their own right, seeing that none of the existing metaphors as entirely satisfying—the emergence of artificial life nevertheless arouses within us a fundamental curiosity. It generates equally disturbing questions and among some people spawns an equally deep-seated intellectual and emotional resistance, as did research in artificial intelligence in its attempt to understand the human mind. If artificial life was simply a collective designation for a few computer models or a new method brought about by the cooperation of several natural-science disciplines, it would hardly have aroused the same virulent reactions. Why the undeniably curious sentiments about artificial life?

It has been said that science demystifies the world. It is closer to the truth to say that science, when it is at its best, opens the world up for us, bringing daily realities under a kind of magic spell and providing the means to see the limitations of what we think we know, and the scope of what we do not at all understand. Even though physics, chemistry, biology, and other sciences have helped to draw a grand map of reality, the blank spots on the map are never completely filled in. The more one draws, the more details must be investigated, and the more new horizons expand. Thus, knowledge has a kind of fractal structure: for every newly collected fragment of knowledge, we reach "farther down"; that is, we can with greater accuracy say more about nature and extend the universe of conversation about it and with it. However, we simultaneously extend our limits to the unknown. And if we restrict ourselves to the fragments, without trying to consolidate them into larger wholes, we succeed only in digging ourselves deeper and deeper into the same hole, and we see only a steadily decreasing part of the horizon. This is the geometry of discovery.

There is more between DNA and the organism than James D. Watson ever dreamed of when he told the world that he and Francis Crick had solved the riddle of life. It was in 1953 that they realized that DNA had to be built up as a double helix. But there is more—not because life is a mystical, vital essence, as previously believed—but because the research within molecular biology that followed Watson and Crick's discovery, has revealed a highly organized system of processes, from the individual cell's microcosm to the total ecosystem. This points toward new questions and new limits for what we can conclude from the analysis of individual chemical compounds. We can understand how DNA specifies the sequence of building blocks that make up the proteins, yet the DNA molecule cannot tell us how the proteins, metabolism, cells, and organ systems in an animal function as a coherent whole.

In the Beginning Were Some Bits

How does a cell emerge? What is it that controls metabolism? How is a hand created? What are the mechanisms of evolution? In order to answer these kinds of questions, it is no longer sufficient to continue with the methodological reduction of complex systems to their individual components, which has been the normal procedure for biology. It is necessary to find methods to describe the *total* system, the cell organization of the organism being analyzed. The structure of the body and of the cell, its dynamics in time and space, and the sudden appearance of new traits at higher levels than those of the component parts (often called "emergence") are quite important themes in the new biology.

It is here that the computer's abilities to manage data and perform calculations on many bits of data also come into play. Artificial life is interesting, not only as a proof that life is easy to create, but also as a broad analogy, a class of models of complex calculated systems that share ecological and evolutionary conditions with many of the real organisms found in nature.

If life is characterized by processing the special form of biological information found in the genetic material, then

THE GAME OF LIFE

life might be viewed as having begun as a kind of natural information-processing system. Life started in the primordial soup, from which originated the large molecules that could store—"remember"—information. This is one of the general theories of life's emergence.[6] In the beginning there was information.

In this sense, it seems plausible that our information-processing machines ought to be able to imitate Nature's trick. But what is information anyway? Is information simply a sequence of bits that allow themselves to be counted and manipulated by a machine or by an organism?

One can define information in several ways: one is as anything that can function as an answer to a question: In the beginning was . . . the answer? Can the answer indeed lie at the beginning? If life is information, and information is an answer, what was the question? Does information not assume a cell, a unit in space, that can interpret and utilize this information in its own interest—a cell that (anthropomorphically speaking) can ask the question, "How do I organize my survival?" and that seeks to interpret the signs it encounters in its environment? The artificial-life approach asks these questions in a radically new way.

A-LIFE AT LOS ALAMOS

In September 1987, 160 computer scientists, physicists, philosophers, biologists, anthropologists, and several other kinds of academics gathered for the first international conference on artificial life, or a-life as it came to be called. The conference took place at the Los Alamos National Laboratory, the sprawling, largely military research center in New Mexico where the atom bomb was built. Today, the mountains of New Mexico are also home to more peaceful pursuits. It is here that the data base for the human genome is being constructed (the complete micromap of human DNA). The a-life conference was sponsored by three organizations: Los Alamos's own "Center for Non-Linear Studies"; the small, but important Santa Fe Institute, long one of the leaders in research on complex systems; and finally, Apple Computers, which in fact delivered the hardware for a large number of artificial life-forms.

During the five intense days of the conference, a myriad of interesting systems was presented. Models for the emergence of life and its evolution were presented as well, including models for selection and adaptation of cooperating "organisms" in simulated ecosystems; self-reproducing automata; flocks of birds and schools of fish; computer plants that grew large and beautiful with the help of simple algorithms (the recipes from which programs are written, in this case formulas for tree-branching and length of growth); and much more. Everything that the heart could desire was here, from the more serious attempts to illustrate and test existing theories of growth and development to pure fooling around. The latter had more in common with arcade computer games and screen patterns in science-fiction films than any serious natural-science research. (The conference's nonexclusionary attitude toward what might be considered the field of artificial life has led some traditional and empirically oriented scientists to deny a-life any kind of scientific relevance or legitimacy. In this way they also avoid the difficulty of having to concern themselves with what it is really all about.)

The conference was organized and led by Chris Langton, who for years has worked with a type of mathematical structure called cellular automata, which are ideal to implement and simulate on computers. (Conway's Life game is in fact an example of such a cellular automaton.) Langton has said that he had long felt frustrated about the fragmentary nature of the field of biological modeling and simulation. For years he plowed his way through the literature in various libraries, data bases, and bookstores in order to obtain an overview of the field, although the field did not exist at all as an independent discipline. Theoretical biology had always been relatively ignored compared with theoretical physics. Biologists are experimentalists; only physicists can with good conscience and full sympathy from their colleagues allow themselves the luxury of being theoreticians. There are, of course, good reasons why this is so—witness biology's slower and subsequent development as a discipline and as a science—but people like Langton had difficulties in finding an environment for their interests.

Ultimately, however, Langton's initiative bore fruit. Not only did the Los Alamos a-life meeting testify to an enormous excitement at the new possibilities of understanding biological-development processes via modeling, of which many of the participants proudly exhibited examples. There also appeared something new: people who had been working in isolation, each tinkering with their own model of, say, cell growth, now met others who had worked with other organisms or at an entirely different level of organization, but had used much the same methods. Langton relates that there gradually emerged a collective feeling about the "essence" of artificial life. It was a vague feeling that did not, in the first round at least, lead to any programmatic declarations or any explicit research program, but that Langton later came to formulate in his book *Artificial Life*.[7] Langton's vision of a new research program, which is likely to become an important part of theoretical biology, contains several elements.

THE SEVEN COMMANDMENTS OF ARTIFICIAL LIFE

Seven central points, when taken together, comprise the vision of artificial life in its strongest, most ambitious form.[8] These apply both to the creation of a new research area and to the technical-scientific realization of new types of systems. The significance of these points will be reiterated in the chapters that follow. In its ambitious version, the concept of artificial life encompasses the following ideas.

1. The biology of the possible. Artificial life does not concern the special wet and carbon-based life as we know it here on

earth, which is the subject of experimental biology. Artificial life deals with *life as it could be*. Since biology is only based on one example, life on earth, it is too empirically limited to help create truly general theories. Here artificial life is a clear and quite decisive supplement. It is not certain that we appear as we do simply because of previously existing earthly materials and the accidental evolutionary sequence. Evolution could rest on much more general organizational laws, but these are laws that we simply do not know yet. Biology today is only the biology of actual life. It must become a biology of any possible life-forms.

2. Synthetic method. Where traditional biological research has placed emphasis on *analyzing* living beings and explaining them in terms of their smallest parts, the artificial-life perspective attempts to *synthesize* life-resembling processes or behavior in computers or other media.

3. Real (artificial) life. Artificial life is the study of humanly created systems that exhibit behavior characteristic of natural, living systems. However, what is it in artificial life that is artificial, in the sense of false, unnatural, or humanly created? That which is "artificial" about life *in silico*—all gadgets and information structures in the form of machines, models, and constructed "organisms"—is not the behavior as such. The behavior, the generalized process, is just as genuine as the behavior exhibited by real-life organisms. No, the "artificial" of artificial life rests solely in the *components* (like the silicon chips, formulas, computational rules, and the like) of which it consists. These are designed by us. The behavior, however, is produced by the artificial life itself.

4. All life is form. Neither actual nor possible life is determined by the matter of which it is constructed. Life is a process, and it is the *form* of this process, not the *matter*, that is the essence of life. One can therefore ignore the material and instead abstract from it the *logic* that governs the process, taking it out of the concrete material form of the life we know. Hence, one can thus achieve the same logic in another material "clothing" or substratum. Life is fundamentally independent of the medium.

These four theses are related. It is precisely because life is a form that the biology of the possible can be studied by a synthetic method in which the form or formal mathematical descriptions (via computer calculations, for example) can be made to achieve a behavioral sequence that is just as authentic as earthly life itself. From here also follow three additional commandments about the way in which artificial life must be constructed:

5. Bottom-up construction. The synthesis of artificial life takes place best via a principle of computer-based information processing called "bottom-up programming": at the bottom many small units and a few simple rules for their internal, purely local interaction are defined. (This is the real programming.) From this interaction arises the coherent "global" behavior at the general level, behavior not previously programmed according to specific rules. Bottom-up programming corresponds to the fact that our proteins are "programmed" relatively explicitly by DNA, but there is no gene that directly specifies the form of the face or the number of fingers. This kind of programming contrasts with the dominant programming principle within artificial intelligence (AI). Here one attempts to construct intelligent machines by means of programs made from the top down: the total behavior is programmed a priori by dividing it into strictly defined subsequences of behavior, which are in turn divided into precise subroutines, smaller subsubroutines, etc., all the way down to the program's own machine code. The bottom-up method in artificial life imitates or simulates processes in nature that organize themselves. We might also call these processes "simulated self-organization."

6. Parallel processing. While information processing in a classical computer takes place sequentially—similar "one-logical-step-at-a-time" thinking is also found in classical AI—the principle for information processing in artificial life is based on a massive parallelism that occurs in real life. In real life the brain's nerve cells work alongside each other without waiting for their neighbor to "finish his work"; in a flock of birds

it is the simultaneity of the many birds' individual small changes in the direction of flight that gives the flock its dynamic character. Artificial neural networks are a typical example of parallel information processing and, hence, a kind of artificial life. (That the parallelism of neural network models can be simulated on sequential computers is simply a stroke of luck and says nothing about the computational principle itself.)

7. Allowance for emergence. The essential feature of artificial life is that it is not predesigned in the same trivial sense as one designs a car or a robot. The most interesting examples of artificial life exhibit "emergent behavior." The word "emergence" is used to designate the fascinating whole that is created when many semisimple units interact with each other in a complex (nonlinear) fashion.[9] In computational terms, it is the bottom-up method that allows for the emergence of new, unforeseen phenomena on the superordinate level, a phenomenon that is crucial for living systems.

An ecological balance in a small lake with plants, plankton, invertebrate animals, and fish is a fine example of dynamic emergent behavior that cannot be explained without the entire process of which it consists. The balance may well contain chaotic aspects (the system will perhaps never completely repeat its own movements). The important aspect, however, is that this dynamism forms an integrated behavioral whole for the entire system; that is, it has a property that does not characterize its individual component elements.[10] It is similar to partly genetically determined embryo development, where one cannot predict the entire system's phenotype (the individual's actual appearance) from the genotype (the specific set of genes), even if the genotype could be completely known. (Often it is the reverse: one deduces the genotype of a particular trait from a knowledge of its phenotype.) We can generalize as follows: in artificial life, the system's P-type cannot be predicted from its G-type, in any case not for any random program.[11] The G-type comprises the simple rules under which the system operates; for instance, the two rules in Conway's "Life" game. The P-type is the model's overall emergent behavior, such as the glider's diagonal wriggling down through the Life grid.

Artificial life is emergent life, but it is not the mystical emergence that the old vitalists dreamed of: artificial life is in a way a mechanist holism.

With artificial life we have a new, creative way of dealing with the contradiction in biology between reductionism and holism. J. Doyne Farmer, who has worked with a model for the emergence of life, believes that artificial life makes it possible to be a mechanist and a vitalist at the same time.[12] Creativity within the field is also expressed linguistically: new expressions are arising almost daily. The researchers who regard themselves as belonging to the community of artificial-life researchers have begun to call themselves "alifers." Not every alifer would wholeheartedly adhere to all seven of the basic ideas listed above. Among alifers there is remarkable interest in the philosophical and ethical questions that the entire initiative raises. This was seen especially at the second international artificial-life meeting, held in Santa Fe, New Mexico, in February 1990, where a central topic of discussion was how life could be defined at all.

Do we really know what life is? We know it intuitively, of course, but this is the same as asking, "What is thinking?" There are psychologists and philosophers who do not regard research in artificial intelligence as real psychology; that is, as at all relevant for understanding the specifically human mode of reasoning. People often think in analogies and images, for example. The AI-based expert systems do not. Artificial intelligence may be seen, rather, as a kind of nonempirical experimental logic, mathematics, or semiotics.[13] In the same way, research in artificial life can be seen as an experimental (and biologically inspired) philosophy that investigates the conditions under which anything can be considered to be alive, to be biological life, b-life. Perhaps it is here that a-life research will have its greatest impact.

Biologists find it difficult to describe the mechanisms of organisms molecule by molecule. It is not much easier for the alifer, who wants to simulate the logic of self-reproduction or other aspects of the game of life. Yet life itself, before being inserted into the scientific framework and systems, seems to contain an imponderable, inexplicable, unutterable, unbelievable, and unapproachable lightness.

Chapter Two

WHAT IS LIFE?

VARIOUS ANIMALS may have an enticing or a repulsive effect. They can fascinate or they can send shivers down the spine. There is a degree of unpredictability connected to this observation. Perhaps this is why animals fascinate us: like us, they can determine for themselves where they will walk, creep, crawl, or swim. This autonomy or independence comes from within. We not only *can* do it ourselves, we *want* to do it ourselves. Even plants and single-celled organisms, though unable to move as we do, have a degree of autonomy.

One criterion for classifying artificial life as genuine life—for determining that it is just as alive as real, natural life—could be this very autonomy, this capacity for self-movement. Inanimate entities like stones or rivers may also move, but they do not decide for themselves how to move. Organisms also have the ability to reproduce themselves, such that the "self" is involved as raw material, process, and product. There is also the capacity for self-organization during the development of the embryo, where the various molecules that are absorbed into the system are structured into a complete organism during a fantastic and until now quite baffling process: that of "a thread growing and taking color, flesh being formed, a beak, wing-tips, eyes, feet coming into view, a yellowish substance which unwinds and turns into intestines; and you have a living creature."[14] If we speak specifically about human beings (and their more or less artificial intelligence) we can, finally, add self-reflection as an extra refinement.

The attempt to create artificial life is nothing less than an attempt to imitate nature's ability to create organic points in physical space; points where a purely physical description no longer captures the essence of the phenomenon, and which must therefore be described with additional concepts and ideas. It is tempting here to view life as locations out of which emerge an autonomous perspective—the organism's own. This perspective is to a certain degree a point of view, even though the creature does not need to be able to see. It may be considered as the point from which the niche is sensed and defined. The more or less self-moving biological unit is equipped with a "subjective" surrounding world (subjective in relation to the constantly limited and species-specific sensing of the environment)[15] and a project that among other things seeks to survive and self-reproduce. Let us therefore hear what biology's parents have said about the physically peculiar, special properties such as self-movement, self-(re)production, self-organization, and other "self-" properties that artificial life seeks to imitate.

The Form and Organization of the Organism

In order to move, to organize (and to understand itself, but this complication we will momentarily ignore), there must first be a "self," a subject, or at least a system that does something "by itself." We can speculate about what this means. The biophilosopher Humberto Maturana speaks of *autopoiesis* (a neologism that literally means self-production) as the fundamental property that characterizes living as opposed to nonliving processes. Here life is meant as "real life." Intuitively speaking, this seems easy enough to understand: the organism continually produces itself when it looks for food, when it eats, when it grows, whenever it seeks to maintain life. It maintains the organization of which it is an embodiment. For Maturana, autopoiesis refers more precisely to a network of production processes in the cell where the components produced (as well as the cell's enzymes) are themselves part of the network and help to maintain it. Autopoiesis is an indivisible quality and cannot be graduated. A system either has this autopoietic quality, and is thereby living, or there is no such system at all. For Maturana, there is no such thing as being half-dead. All products of human design are *heteropoietic*—created from without, they cannot be autopoietic. Hence, Maturana would hardly acknowledge artificial life as genuine life.[16]

Maturana upholds a tradition in biology that promotes a search for the unity of all living things. It is a tradition stretching back to the master thinker Aristotle (384–322 B.C.). Aristotle was one of the first to combine wet, field biology with daring cosmological thinking. He spoke of the *psyche*, or life force, of living beings, some inseparable principle of movement in the body of both animals and plants, something from which life itself had to be understood. Of course, in Aristotle's system physical elements like earth, fire, and water also had principles of movement; the entire Aristotelian universe was subordinated to the same principles, including the idea of purpose. However, Aristotle emphasized that even though a stone has a natural tendency to fall to earth, it is not able to follow this tendency on its own. Living beings do not share only these natural tendencies toward movement and change; they are also able to follow these tendencies on their own power, "for when an animal is in movement, the movement stems from the animal itself."[17]

For Aristotle, the psyche is the *form* of the living organism, regardless of what kind of beast is being discussed. Form is one of the four types of causes he investigated, and these causes ought to be viewed as principles of understanding. Among living beings, Aristotle saw in the psyche an interesting intersection of three of the four causal types: the psyche is both the moving, dynamic principle (the efficient cause) and the purpose (the final cause). And as stated, the psyche is also the bodily essence (the form or formal cause). For Aristotle there existed no Christian dualism between body and soul. He viewed the psyche as the source of several basic expressions of life, such as consumption, growth, reproduction, sensation, desire, etc. The psyche was the central force in the organism's capacity for self-movement and for all similar "self-" functions generally.

It may appear somewhat drastic to accord the quality of psyche to a slime fungus or a dandelion. Our justifiable suspicion is nothing new: after the Renaissance and the break with medieval philosophy, large portions of Aristotle's teachings were considered hopelessly vitalist and strained. Yet we might add that after Aristotle, the concept of psyche took on a new, more anthropocentric meaning. Aristotle himself gave the concept more nuance, stating that plants, animals, and humans each have progressively greater differentiated psyches or powers of the soul.[18]

Nevertheless, a break had to be made with Aristotelian thinking. In the Middle Ages, classical thought had hardened into a stiff system of categories emphasizing the coherence of nature: nature was viewed within the Great Chain of Being, emanating from God and the archangels, down to the angels and humans, and finally through animals, plants, and minerals. The chain was broken by the new natural science. With physics in the forefront, the signal was given for a new mechanistic interpretation of the material world and a more instrumental, interactive relationship with nature.

Many natural philosophers (from the time when natural philosophy was a science) had already attempted to classify the world's inventory of things and creatures into three great realms within the old system: the distinct kingdoms of animals, vegetables, and minerals. In the Renaissance, however, they began to discover commonalities between animals and plants that transcended the borders of their respective kingdoms. The science of biology slowly emerged, making us conscious of the differences between the organic and inorganic universes. This consciousness raising was in latent contradiction to the mechanistic worldview, which insisted upon all things having the general character of machines. At the end of the 1700s, some anatomists, botanists, and natural historians—among them Louis Daubenton—promulgated the idea that the unique feature that distinguished the living from the inorganic was *organization*, possessed only by living things.[19] This difference, Daubenton

asserted, was more fundamental than the difference between plants and animals. Even though animals might appear to be more organized than plants, the life functions of
plants also demanded a high degree of organization, in-
comparably more complex than the beautiful but simple
geometry of minerals. Hence, the image of nature as three
distinct kingdoms was modified by the exploration of features common to all that is living. At that time, the very
term "biology" did not yet exist. The label for this new effort arose in 1800, and the notion of a unified science of life
was first seriously disseminated in the works of August
Comte in the mid-nineteenth century.

Comte is known as the father of sociology because he
sought to establish an independent science of human social
life, a science that could not be reduced to either psychology or biology. The social was a domain in itself. In the
same fashion, Comte also rediscovered the biological as a
distinct and autonomous level of complexity. Even though
the laws of physics could explain much, Comte believed
that the special features of organisms' anatomy, physiology, and development required independent biological
recognition.

A Historical Cocktail

We thus see that the idea—whether considered outlandish
or brilliant—of creating artificial life is predicated on a notion of a unified biology that can be traced back to Aristotle.
This is the notion of "the living" as something generic, a
property of all life-forms: hence it requires a coherent and
comprehensive scientific inquiry. Yet the idea of biological
unity has roots in the concept of organization among natural historians, a concept that rapidly became quite well pop-
ularized, and in the machine thinking that can be traced
back to the scientific revolution in which Descartes went so
far as to regard his own body as an advanced machine. Finally, artificial life requires the computer, a machine that
can automatically compute and calculate, and that has its

own historical antecedents, among them Leibniz's idea of thought as a kind of calculation process. In this sense, artificial life must be understood as information processing. If life is form, it is forms of communication: life is information.

In the spirit of artificial life, therefore, we can mix this new research discipline as a kind of cocktail: taking some Aristotle and some Descartes, and mixing their ideas with the thought of Daubenton and Leibniz. If we add a bit of Comte—i.e., his interest in levels of varying complexity— we have our cocktail. If this cocktail is too sour for our taste we can sprinkle a bit of Darwin over it: Darwin adds the biological realism, evolutionary perspective, and processes such as randomness, selection, and adaptation.

Our selective, wormlike excavation of the intellectual roots of artificial life lays open many philosophical questions. These questions are lively topics of discussion among alifers, and their debate is part of a natural attempt to establish a viable, independent research tradition, a new paradigm.

OTHER BIOLOGIES: EXOBIOLOGY

Life on earth might well have appeared completely differently if evolution had taken a different course. The character of genetic assignment is purely a lottery, and the multiplicity of evolutionary pathways makes it conceivable that instead of focusing on the development of intelligent, social, and city-creating primates belonging to the family of mammals, we could conceive of the emergence of intelligent beings who neither resembled us, nor the other apes, nor any mammals at all, but that instead laid eggs, lived in salty lakes, and evolved completely different types of language, creativity, and art than those with which we are familiar. If developments in the very early period of the evolution of life had taken place under other conditions, the organisms that exist today would be radically different.

On the other hand, the phenomenon of "convergent evolution"—when two completely different evolutionary lines seemingly merge into the same path—points to certain limitations in the conceivable forms of life. Marsupials and placental mammals have independently evolved several species whose ecological niches resemble each other to a high degree. The "normal" flying squirrel resembles a marsupial flying squirrel, the flying phalanger, and the common mole resembles the marsupial mole very much (even though the flying phalanger and the marsupial mole are more closely related to each other than to their respective 'ecological' counterparts among the placental mammals).

This is an example of how the environment provides a framework for the vagaries of evolution: not everything is possible under evolution, and to avoid dying out in the evolutionary play, one must choose among a limited number of roles in the ecological theater. Besides, the bodily form puts certain historically conditioned constraints on the forms that could appear in the future. Evolution does not take place according to a globally optimal design, but by the principle of tinkering. The constraints are also indicated by the relatively small number of basic building plans found for the large classes of animals and plants now living.

When the multicellular invertebrate animals came onto the scene, ecological roles were not so fixed. Evidence of a lost multiplicity of life-forms comes from the paleontological studies of the Burgess Shale fauna in Canada. Stephen Jay Gould pointed out how difficult it was for traditional paleontology to comprehend, or even see, this diversity of animal-construction plans. The unconscious tendency was to impose the established order of classifications on the strangest creatures excavated and reconstructed. A prehistoric monster like *Hallucigenia* is an example of this seemingly maladaptive creativity in the early evolution of multicellular organisms (see fig. 2.1).[20]

Life on earth could have been different, more fun perhaps, if these monsters had survived. Meanwhile, life is certainly different on other planets. At one time it was considered heresy to express the idea that the universe's nearly infinite dimensions could contain conceivable worlds other than ours. Giordano Bruno, in 1600, was burned at the stake

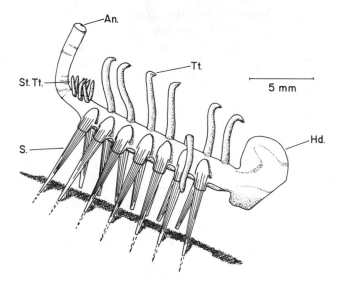

FIGURE 2.1. Hallucigenia is one of the small "prehistoric monsters," that paleontologists have found in Canada's Burgess Shale quarries. Following Gould's description of the animals, and of the entire affair in connection with their (mis)interpretation, the site has become a veritable tourist attraction (see n. 20).

during the Inquisition because he asserted that the universe is infinite and contains a multiplicity of worlds that resemble our own. Modern astrophysical cosmology rejects Bruno's idea of the infinity of the universe, even though it uses arguments other than those of the Inquisition. Yet with regard to the existence of other worlds within a finite universe, astrophysics supports Bruno's contention, supported by a series of assertions partly based on probability: it is highly probable that life emerged on distant planets where the conditions have been just as fertile as those here on earth, even though we do not know the range of possibility—the spectrum of physical conditions—that allow for the emergence of life. For example, no one knows for certain whether carbon and water are absolutely necessary for life, even though we ourselves find it difficult to imagine life without water. To be far from life without water is pure death, we think. We know only our own carbon-based life: carbon, together with hydrogen, oxygen, and nitrogen are the basic elements found in all the macromolecules that compose a cell. It is easy for us to become "carbon chauvinists," capable of dismissing out of hand the potential existence of any other life-forms. This prejudice, say Bruno's modern successors, is without justification in light of the relevant evidence or scientific theories.

Hence, there is a certain logic in researching exobiology—the study of life beyond earth. And once life has emerged, intelligence may follow. Where life has emerged for the first time on a planet, there has commenced—at least in our own earthly eyes—an evolutionary process that could produce thinking beings who might create their own civilization. In the 1960s U.S. research projects on the exploration of the other planets in our solar system sought to investigate whether traces of life could be found there. Even though life on the moon is grossly improbable, in 1969 investigations for traces of organic compounds were carried out on the Apollo 11 moon samples. The results produced no definitive conclusions. Other projects were also begun, such as SETI (Search for Extraterrestrial Intelligence), the goal of which was to search for the existence of alien intelligent beings and to explore the possibilities of communicating with them.[21]

Exobiology became relatively well established, even though many specialists considered it to be a bit outlandish: exobiology was the only empirical science that had no object. Carl Sagan, one of exobiology's leading lights in the 1960s, turned this to his advantage. Even though we may never find extraterrestrial life (other than in films like "E. T."), the exobiological perspective has profound value. Exobiological experiments can be done with other "chemistries" in the laboratory, that is, without the organic materials found in the cells here on earth. This type of experimentation compels us to investigate the validity of the assumptions we normally make regarding the ecology and physiology of life.

One problem that arose with projects like exobiology and SETI was that there was no definition of life. Since we know only about life on our own planet, we do not know which aspects of living systems are basic and which are purely superficial and contingent—that is, aspects that by coincidence have emerged with earth life-forms. An ecologist who knows only about life at the bottom of an ocean would be in the same situation if forced to make statements about life in a beech-tree forest. Therefore, even a single example of extraterrestrial life, regardless of how simple, would have revolutionary consequences for biological theory. This was Sagan's simple argument for exobiology's scientific significance.[22] This same argument is now used by Chris Langton to justify the study of artificial life.[23] Artificial life is precisely "the biology of the possible."

If we search for life elsewhere in the universe, we must know what it is we are looking for: what is required for us to call, for example, certain specific chemical processes on Mars "life"? Or, seen in the eyes of an immigrant: how would a Martian who landed in the Sahara be certain that the sand's bacteria or the nomads' camels were alive?

How should we respond to the alifers, hackers, and computer freaks who in a few years will assert that they have created genuine computer life that functions at the level of horses and pigs? Pigs are clever and sensitive, say those who are friends of animals (and presumably, the pigs themselves). Will alifers then demand that animal-protection regulations also be applied to artificial life as well? What kinds of rights do artificial life-forms have? Some people believe that all organisms have a certain right to exist. Does this also apply to computer organisms, or is this thought simply too absurd? And why not? Here there is a need for precise definitions of life, much as jurists today need definitions for different types of genetically manipulated species that have been submitted for patents. The future always offers surprises. Twenty years ago we thought that issuing a patent on genetically designed species was an absurd fiction. Today it is a reality, and a jury always shows itself adept at handling absurdities in practice (which does not yet rule out the absurd). Let us examine some of these vitally important attempts at definition.

DEFINE YOUR LIFE!

In 1878, the physiologist Claude Bernard cited five generally common features of living things: organization, generation (reproduction), nutrition, development, and susceptibility to illness and death.[24] Numerous other researchers have since attempted to characterize life using several of its properties, chemical as well as biological. Louis Pasteur pointed to the fact that optic activity was intimately connected with life.[25] We know today that a cell's amino acids are optically active: in solution they can rotate incoming plane-polarized light to the left. However, we do not know whether this feature is universal, that life must necessarily be "left-rotating," and we have hardly any reason to believe so. This is certainly not the case for computer-borne lifeforms, which are neither optically active nor chemical in this sense.

In the middle of this century, biology received important inspiration from physicists such as Niels Bohr, Max Delbrück, and Erwin Schrödinger. Their ideas helped to form a concept of life based on "code-script" (to use Schrödinger's expression from 1943), which Watson and Crick, in 1953, came to identify as the particular structure of the DNA molecule, the molecule containing the "universal" genetic code for protein molecules.[26] It was not enough, however, to clear up the question of which minimal set of properties are necessary and sufficient in order for a physical system to be considered alive. One could certainly envision entirely different kinds of information-bearing molecules that could function as a genetic code.

Today, each subdivision of biology has its own way of

defining—and thereby viewing—life. There is no accepted agreement on what biology studies, and a whole range of different definitions exists in the various branches of present-day biology.[27]

The *physiological definition*, for example, would be as follows: every system that can carry out functions such as ingestion, metabolism, excretion, breathing, movement, growth, reproduction, and reaction to external stimuli is a living system. Even though the criteria are meaningful, they are not logically satisfying. All of them can be applied to systems that no one would call living, such as automobiles. A car is, after all, auto-mobile (self-moving), it can be said to "eat" gasoline, excrete fumes, etc. Moreover, some of the above criteria are not fulfilled by certain organisms that clearly we would call living, e.g., bacteria that do not breathe oxygen. The above criteria are unable to incorporate the nearly intuitive knowledge we have that makes it evident to us what life is and what it is not.

The *metabolic definition*: a living system is one that is distinct from its external environment and that exchanges materials with its surroundings (i.e., has a metabolism) without changing its general properties. Here again there are exceptions. A seed may have been in its dormant stage for centuries without any noticeable metabolism, yet it lives. A whirlpool in a river is quite unbiological, yet while it spins it retains its organization and exchanges materials and energy with its surroundings. The whirlpool certainly does not have the sharp delimitation of a cell membrane, but the membrane, too, is "veiled" if viewed from a suitably large scale. (This hints that delimitation, far from being an objective trait, is drawn by the organism itself; it is defined both from within as well as by a physical barrier. This is implicit in the concept of autopoiesis.)

The *biochemical definition*: biochemistry and molecular biology see a living organism as a system that contains reproducible, genetically transmissible information coded in DNA and RNA, and that governs the creation of proteins, among them the enzymes that help catalyze the various metabolic processes. A counterexample could be prions (proteinaceous infectious particles that cause scrapie)—i.e., simple, virus-like organisms whose genetic material does not seem to be based on nucleic acids such as DNA—or the DNA-less artificial life that we have already described. The biochemical definition therefore suffers from being "carbon chauvinist."

The *genetic definition*, which is also an evolutionary one, views life at the level of the population: life exists when a system consists of certain units (organisms) that can reproduce and transmit their genes (the genetic instructions about the creation of the unit, the phenotype) to a successive generation. The system must also be capable of mutation, so that new information can arise and certain instructions can be altered. If an altered instruction allows the unit to adapt better than others of its species, that instruction will probably be maintained in the population. (The mixing of information is important: genes meet and some sweet recombinations may emerge, it is indeed a matter of good chemistry.) Often, however, the genetic changes will worsen the organism's chances for survival and reproduction. Only in exceptional cases will the units created be significantly better adapted. These will have a better chance of survival and their type will thus spread within the population; this is natural selection. The phenomenon allows for development in the direction of increasingly more refined adaptability and a higher level of integration. In this Darwinian perspective, complex organisms emerge due to replication, mutation, and replication of mutations. In sum, a system that can evolve by natural selection is a living system.

That we can use artificial means to construct systems that evolve by natural selection poses no special problem for this definition. These systems are also living, says Carl Sagan.

The *thermodynamic definition*: From the perspective of thermodynamics, life is an open system that exchanges energy and materials with its environment. In this case, the crucial Second Law of Thermodynamics does not seem to operate. The Second Law states that in an isolated, energetically closed system, the system's tendency toward disorder—entropy—has a tendency to grow. As disorder increases, the system's structure and organization disappears. Disorder is incompatible with life. For open systems, fortunately, it is not impossible that entropy can decrease; that is, order can be increased locally (in the organism) while decreasing within the total system (organism plus environment). This is precisely what is typical of living systems: they have the ability to create order out of chaos simply by being assured access to a continuous flow of energy through the system (energy with low entropy as input and energy with high entropy—heat, for instance—as output). An amoeba "eats" order—that is, it ingests materials with more order (energetically speaking), which it utilizes to maintain and rebuild its own organization, and it expends heat (spreading disordered energy) to its environment. An entire ecosystem operates on the same principle: the ordered input consists of light from the sun, the disordered output is heat to the environment, ultimately to outer space.

Like the other definitions, however, the thermodynamic perspective is weakened by the fact that other, intuitively nonliving systems also function according to the same principles. Rather, we might instead classify life as a partial set of thermodynamically open systems. We know of innumerable other examples of "organic-like" cyclical processes in open, though purely physical, systems that can also create ordered patterns from flows of energy. One example is the Bernard cells created by pouring oil and chili powder in a frying pan and then turning on the heat. The powder follows the "cells" of circular liquid flows in the oil caused by heat convection; the powder changes from being disordered and randomly distributed in the oil and produces beehive-like patterns. Here it is the self-organization of a "dead" system that shows similarities to life (or to order in any case) without our being able to define our powder-oil–frying pan system as "living." The exploration of such systems—including places other than the kitchen—that are far from (thermodynamic) equilibrium because of the energy exchange with the environment has contributed to softening up the boundaries between physics and biology, without, however, leading to adequate explanations of the specifics of biological nonequilibrium systems.[28]

With all due respect to Maturana's "either-or" idea of life, perhaps we may have "semiliving" systems, organisms that are neither fully autopoietic nor completely without the ability to reproduce their own components at least partially. Presumably, life emerged in a gradual, long-term process whereby primitive proto-organisms were created in the teeming, primal soup of the seas. Life is still not completed; it is a partly indefinite phenomenon of becoming, not a completely determined state of being. To use a physical analogy, we might say that life can be viewed as a "critical" phenomenon, as a phase transition that can occur quite suddenly or in a limited zone of the phase space from which complexity can emerge—that is, between crystalline order and random disorder.[29]

Contemporary biology has accepted that no single definition of life will be without exceptions or other weaknesses. But the five definitions, combined according to need, are adequate to cover most of the earthly forms we know of. An organism like a turnip or the human body has all of the above-mentioned properties. Yet theoretical biology is forced to accept that life is a vague, cluster-like concept (a prototypical concept). Every property we accord to life is either so broad that it also applies to many nonliving systems, or too specific, so that it cannot encompass counter-examples that we intuitively consider to be living.

Since we are all born carbon chauvinists, we would be insecure about life in media other than in the organic-chemical one in which we live. The genetic definition is in one way the least chauvinist or material bound. It certainly describes the abstract evolution of information-processing systems via natural selection. It does not base itself on the organism as such, but on the population or on the phylogenetic lineage, the "family line": to have a historical, natural line going backwards is an important feature of life in the biological sense.

Let us now examine a list of the key properties of life, real life, as elaborated by J. Doyne Farmer and Aletta d'A. Belin (see n. 12). Farmer and Belin accord life the following characteristics:

1. Life is a pattern in space/time (rather than a specific material object). In other words, life is a distinct form of organization. We are, after all (and fortunately), more than what we eat. The molecules in our bodies and the cells in our tissues are renewed and exchanged innumerable times during our lifetimes.
2. Life loves self-reproduction (even mules, which are sterile, are created via a reproductive process).
3. Life is associated with information storage of a self-representation; that is, a partial description of itself (or of certain components necessary for production of the remainder under the system's continual self-organization).
4. Life thrives with the aid of metabolism; see the metabolic definition above.
5. Life enters into functional interactions with the environment. (That is, organisms can adapt, but they can also create and control their respective local environments.) Organisms have the ability to selectively respond to external stimuli (what the old physiologists called "irritability").
6. Parts of living things have a critical internal dependency on each other (which means that organisms can die).
7. Life exhibits a dynamic stability in the face of perturbations (it can maintain form and organization up to a certain limit).
8. Life, not the individual but its lineage, has the ability to evolve (see the genetic definition).

The fact that life appears to be a conglomeration of properties points toward what we might call a metaproperty in the original biological view of life, the view that stems from the classical natural-historical notion of life as an especially high degree of organization: rather than being an all-or-nothing concept, as Maturana's autopoiesis, life has continuum-like features: it is a continuum property of organizational patterns, some of which are "more alive" than others. Even a property such as natural selection, which can be formulated in general terms (regardless of whether selection occurs for particular genes, organisms, or species, and regardless of whether we are speaking of selection in biochemically or silicon-based systems), is not necessarily inevitable as the only possible mechanism of evolution. Evolutionary biologists have discussed other mechanisms, such as Sewall Wright's "random drift." In a logical sense, life must be considered to be a fundamentally vague concept that reflects a genuine vagueness, a continuum in nature.[30]

Metaphorical Slippage

The criteria for the phenomena we call life must therefore be used delicately when we investigate whether life can thrive in the cool chips of the computer or in entirely different media, as Farmer, Belin, Ray, and many other alifers assert.

Research on artificial life uses programs that achieve lifelike properties and can be considered, at least metaphorically, as living. Thomas Ray's "organisms" and Conway's Life game, which can evolve not just gliders but several more complicated structures (see figs. 3.6 and 4.3), are examples. Ray's organisms are deliberately designed to move around in an artificial environment simulated inside the computer. The infamous computer virus is another example of a self-reproducing program. Let us first dissect this scientific metaphor and then examine whether computer life fulfills the eight criteria for life set forth by Farmer and Belin.

Just as organic life consumes energy that comes from the sun in order to organize material, digital life can be seen as a consumer of the main computational device in the computer, the central processing unit or CPU. The CPU is used during a certain period of time to organize the memory.

Ray directly compares CPU-time with the energy resources of real animals, and allows computer memory to correspond to the spatial resource.[31] And just as organic life evolves via natural selection, digital life can be seen as evolving via competition among algorithms, i.e., the procedures or simple instructions that form the basis of a program. Algorithms compete for the machine's CPU-time and for space in the memory-storage area. Only those programs that most effectively utilize the resources (and eventually utilize each other, as Ray discovered for parasite programs) will be selected; these will create the basis for new generations of programs. While the memory, the CPU, and the computer's operating system can, in ecological terms, be viewed as parts of the "abiotic environment," Ray's self-replicating programs in assembler code correspond to the ecosystem's organisms.

In computer science, assembler languages consist of names that refer to an especially useful group of instructions in machine language. The machine language is the computer's most basic set of instructions, its computational "natural" language. Where the machine-language word for "addition" may be "010111000," the assembler language may render this instruction simply as "ADD," much like instead of tediously specifying a DNA molecule atom by atom, one can be content with indicating a sequence of abbreviations for the bases (A, G, C, and T) of which the molecule consists. When an assembler instruction is to be executed, it is automatically translated into machine code for execution in the CPU according to the CPU's own set of instructions and the operating system itself.

If life is viewed as a process that can be both chemical and computational, it is tempting to use the computer's machine instructions as the artificial chemistry that forms the basis for those organisms designed to live in the machine. While the metaphors between language and life can always be construed in various ways,[32] Ray views the instructions as corresponding to the cell's amino acids (building blocks in the proteins) because these instructions are what he calls "chemically active." They manipulate bits, bytes, CPU registers, and other basic components during the execution of their program. Basically they consist of a chain of machine instructions. The proto-organism that Ray released in his own computational soup had eighty such instructions, measured in assembler code. But Ray's digital organisms can also be compared to the RNA molecules of the primal soup that existed four billion years ago, at the time when the first genuine life-forms emerged. It is thought that the RNA molecules were information bearing and exhibited metabolic activity.

If the digital organisms were to be such successful species that they could evolve inside their owners' computers, could they not get out of control? Could they not by themselves lead to the creation of computer viruses and "worms," which could unintentionally spread into the giant national and international network and constitute a real threat to the great data systems? Researchers within artificial life view this as a real danger. They therefore avoid allowing the computer's own CPU and memory to constitute the environment in which the digital organisms are permitted to play in. Instead, they use the computer's ability to simulate other computers. On a normal, powerful microcomputer, a program can simulate another, so-called virtual computer with it own "virtual operating system." It is the virtual computer's computational time and memory that is made into the "environment" for the artificial organism. Hence, Ray's virtual computer, which he calls the Tierra Simulator, consists of nothing other than a program and data in his actual computer. The virtual computer is just one among many other programs and data, and it is therefore no greater threat than a normal word-processing program or data system (see fig. 2.2).

Since the virtual computer is a computerized container within the real computer, intended to insure against the escape of artificial life, one can term this solution "computational containment." It corresponds to the attempts made by gene splicers to develop "biological containment" that

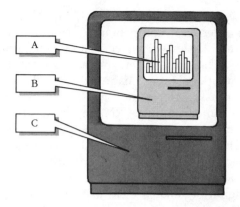

FIGURE 2.2. A strongly schematic depiction of the principle of computational containment. The computer, marked C, is the real machine that carries out the simulation of B, the virtual computer, which in turn constitutes the environment of the computer organism, A, here represented in a block diagram showing the number of various kinds of organisms at a given point in time. (In practice, B is not visualized as shown here.)

would insure that any gene-spliced organism that escaped could not survive (via, for example, "suicide genes").

Computational containment permits other obvious advantages. Ray's artificial universe is not bound to a specific type of hardware and can easily be transported to newer computers. His "universe" can be transmitted by modem or transferred onto a diskette and sent through the mail. Finally, the machine languages of real computers are not designed to support any evolutionary structure. Most machine languages are extremely sensitive to mistakes in programs. Not much is needed before the machine goes down; this is because only an extremely small part of the huge number of possible combinations of instructions can in fact operate as efficient programs. If the real machine code is exposed to mutation and recombination, the program cannot execute its instructions as intended. It will not be functional.

The Danish physicist Benny Lautrup has therefore proposed that we distinguish between two types of computer organisms: "real" and "virtual."[33] The virtual computer organisms are those designed to crawl around in games, in cellular automata, or in virtual environments like the Tierra Simulator. Their existence is completely dependent on people having consciously designed a specific habitat for them inside the machine.

Real computer organisms are also designed by humans (as are all known computer viruses). However, once a real computer organism escapes from its creator, it is not subordinated to the same form of control as a virtual organism. Such viruses function as enemy codes, and their environments are real (not virtual) computers, real hardware, including the diskettes we innocently exchange with each other, where they sit for much of their time and wait to be transferred to another hard disk, a mainframe, or a network and then wreak havoc. From here also come worms, independent programs that multiply and cause damage, as well as "bacteria," programs that replicate themselves until all informational resources are exhausted. Real computer organisms are those that are the most autonomous, beyond the fine control of humans, and able to act as parasites on human instruments. It is quite improbable that virtual organisms can escape computational containment and become real ones.

ARE THEY REALLY ALIVE?

The question of whether computer organisms are life is still unresolved. However, if we reexamine Farmer and Belin's eight criteria one at a time, various matters can be cleared up.

1. Both computer viruses and Ray's parasites are patterns; they are information structures rather than material objects. (Whether they are so for the same reasons that biological life consists of "forms" is a question that must, however, remain open.)

2. The digital organisms and viruses are able to reproduce

themselves (though only by utilizing an "analog," a material machine that can execute this operation in a purely physical mode).

3. They also have self-representation (which is somewhat special in that it nearly overlaps with the organism as such, just as the biological virus often exists only as an RNA- or DNA-string).

4. They have in a certain sense a metabolism, inasmuch as the machine instructions can be interpreted as Ray has, not just as logical primitives, but as chemically active entities (though with a radically different kind of chemistry than the cell's enzymatic activity).[34] In any case, the instructions conduct and redistribute some of the computer's electrical energy during the creation of heat in order to maintain themselves and to respond to specific "stimuli."

5. In this way the computer organisms enter into functional interactions with their environment, whether artificial-virtual or "real" hardware.

6. And inasmuch as digital organisms are small functional wholes, we must acknowledge that their parts are mutually interdependent in a critical way. Ray's organisms can certainly die.

7. However, they are also stable structures within their preferred environment, even though one could discuss how dynamic these programs are.

8. And after having seen the various forms of programs which emerged within Tierra after the release of the eighty-instruction-long ancestor organism, they must clearly be accorded the ability to evolve in a "family lineage." Ray has even attempted to see his simulations in terms of macroevolutionary patterns corresponding to the patterns that real paleontologists have described in the geological fossil record; such patterns would entail the stability of species over long periods of time, punctuated by rapid jumps to new life-forms (see also fig. 2.3).

It appears as if artificial life (both real and virtual) is genuine life according to all the above criteria. The problem lies

CRITERIA OF LIFE:	REAL COMPUTER ORGANISMS	VIRTUAL COMPUTER ORGANISMS
1. pattern in time/space	+	+
2. self-reproduction	(+)	(+)
3. self-representation	+	+
4. metabolism	(+)	(+)
5. organism/environment relation	+	+
6. parts in a functional coherence	+	+
7. dynamic stability	(+)	(+)
8. evolution	–	+
9. autonomy	+	–

FIGURE 2.3. Scheme of the criteria for life. Autonomy is here understood as defining the difference between real and virtual computer organisms. The individual criteria can be debated further, of course. A + sign denotes the criterion being fulfilled, a (+) sign as possibly fulfilled, and a – sign as unfulfilled.

with the reservations we must make along the way. We are speaking of digital organisms whose self-reproduction is entirely informational, and in a certain sense formal. It is not a case of material growth and consumption of matter as with biological organisms. A biochemist would hardly be able to acknowledge the changing electromagnetic states in a computer's hardware as metabolism. The cell's metabolic network certainly looks quite different. Real organisms are both physical and linguistic; they contain both material substance and signs in their DNA. An amoeba, for instance, is both analog and digital, its digital "program" interacts with a real external world via the "analog" amoeba, i.e., the material structure that makes up the cell body. The character of such a biophysical structure is itself decisive for the kind of differences in the environment that make a difference for the organism.[35] The purely digital organisms are very unbodylike.

What about real viruses then? HIV and other retroviruses

are also only "digital." The analogy is indeed stronger here, but it is rather because it is doubtful whether viruses can at all be considered organisms. Unlike the amoeba, viruses have no independent metabolism; they have only the capacity for (pseudo-)self-reproduction by exploiting the metabolism of the host cell. Moreover, real viruses have a specifically three-dimensional physical structure that is significant for their biological functions. In this sense, too, they are not only formal or informational.

Farmer also admits that computer viruses are not stable during major electrical disturbances, and that they are not able to evolve via natural selection. They do not belong to a well-defined phylogenetic lineage. Genuinely robust computer viruses must be able to tolerate changes in their programs. And if they are to resemble biological viruses, they must evolve by themselves, without the involvement of their human programmers. However, it may just be a question of time and algorithmic inventiveness, and perhaps a bit of computer cynicism, to accord these properties to the known computer viruses.

Despite these reservations, Farmer and Belin see computer viruses as a sign that life-forms have already begun to thrive in media other than the biological. It is possible that these alien phenomena do not manifest the entire continuum of bio-organizational properties, as we have discussed. But these creatures are also only the beginning.

Chapter Three

THE LOGIC OF SELF-REPRODUCTION

LIVING ORGANISMS multiply, they reproduce themselves. If we put aside the emergence of life on earth, new life is always created from already existing life. The conditions are no longer the same as they were when life first arose. Life now comes from life, or as Rudolf Virchow formulated it almost 140 years ago, "All cells are created out of already existing cells" ("Omnis cellula e cella," 1855).

Evolution's most successful theme is its emphasis on variations. Reproduction can be asexual and solo, as with the amoebae's division, the plant's generation of offshoots, swollen roots, continuation bulbs, bulblets, or brood buds. Animals can in the same fashion reproduce via simple division (certain kinds of starfish), by "strangulation" constrictions (among bristle worms and medusas), by simple budding (the freshwater polyp), through parthenogenesis or virgin births (greenflies, aphis), and innumerable other variant methods. Or they can reproduce collectively as with the creation of colonies. Among most species, however, bees as well as flowers, production of offspring takes place via sexual reproduction, using the clear advantages of genetic recombination. Sex establishes a genuinely common pool of genetic possibilities in the population that can be useful for future flexibility. (It is sexual reproduction that defines a group of organisms as being a biological population and not just a collection of individuals with common descent.) Sexual reproduction is a very social activity, marked with infinite refinements in sexual behavior, from the animal-pairing equipment of rutting songs, calls, ritualized battles, and complicated foreplay and afterplay, to the pure cannibalism of certain types of spiders. Perverse? It is not the spider who is perverse, but nature that is polymorphous. Volumes upon volumes of natural history are filled with intensively detailed studies of these fascinating processes of procreation. Such descriptions once constituted a highly appreciated popular science, especially in the Victorian period. In order to avoid drowning in these details, let us focus on the kinds of biological generalizations that can be made with regard to reproduction. In particular, what is the essence of self-reproduction?

DIVIDE OR DIE

First, we can acknowledge that asexual fertilization is based on cell division and growth, which among multicellular organisms is often associated with specialized sexual organs. Sexual reproduction resembles the reproduction of genetic dialogue between two organisms. Technically speaking, it demands a previous reduction division or meiosis, in which the number of chromosomes per cell is halved in the formation of the sex cell, the gamete (without meiosis the genetic material would double for each generation). Then comes the fusion of egg- and sperm-cells (which again gives a double chromosome number); and finally, the form-creating growth (morphogenesis) and differentiation during the subsequent creation of a new individual.

However, if we ignore the diversity of specific modes of reproduction and species-specific life-history variations, self-reproduction is basically a process whereby cells divide. They divide, of course, in a way that maintains their organization, identity, and relative autonomy as cells. Even the simplest single-celled organisms (amoebae, bacteria, flagellates, algae) must maintain this complexity. The cell's most complicated task is to reproduce itself as a system. It must divide or die. As a material lump consisting of individual molecules, it will always perish. As a formed whole, however, it can be maintained via division and renewed growth.

Living systems, like all systems, always have a certain kind of output: life produces something, and it can be anything at all. However, a system's production *of itself*, derived from an input of raw materials, is something absolutely essential (and perhaps quite special) to life, a part of the logic of life.

THE LOGIC OF LIFE—AND OF THE MACHINE

The idea of the logic of life operates in biology's attempts to determine the laws, rules, or customs to which all life is subject. Contemporary biology attempts to describe the logic of life. Researchers in artificial life also make the same attempt. Do they find the same logic?

One exposition of the research program lies in extension of the mechanistic tradition in biology: artificial life is possible precisely because living *organisms* themselves are types of *machines* that can reproduce themselves. And since a machine's functions—its logic—can in principle always be imitated in new constructions (whether with the same or other materials is unimportant), it follows that life itself, the organic machines, can be constructed. The attempt to construct artificial life is therefore also an effort to find the logic of biological machines. This search is scientific and legitimate, regardless of whether the method is the analysis of the living things found in nature or the synthesizing of new ones.

This argument builds on a kind of analogy between machines and organisms. The problem, however, is that unlike organisms (momentarily leaving out artificial ones), machines are products of human design. The organism and the machine are not identical concepts in language, neither in the content or range of the concepts, nor with respect to the way they are used in everyday speech.

It is only when we subject the organism to the evaluations of its human designers—how functional is this creature? Can its limbs support it optimally?—that the instrumental-machine metaphor becomes tempting and falls into place quite naturally.[36] The instrumental view lies to some degree within the word *organism* itself: a system of organs, a whole composed of parts, where each part is a functional tool related to the other parts and to the whole. While "organism" today creates associations to something holistic, soft, green, and vibrant, the root of the word lies with the original Greek meaning: instrument, as in machine or mechanism.[37] Of course, the original semantic overlap in the etymological roots of the words "machine" and "organism" should not hide the fact that we are now speaking of two different concepts, each with its own meaning and area of application. Therefore, whenever we refer to living beings as machines, we are using a metaphor.

In science, metaphors can be reformulated into more precise, well-defined concepts via generalization. One might say that both the humanly created and the biological machines that have arisen by natural selection are two instances of a generalized machine concept, understood as any physical structure whose causally determined function can be described via Newtonian laws of mechanics. The broad concept of machine can be extended to the entire world—the Newtonian universe was perceived as just such a giant clockwork. The problem for biology is that this concept does not account for the unique ability of animals and plants to reproduce. In a certain sense, the mechanistic philosophy describes only "the death" in the machine—the movements of its inorganic parts—but life, or "living feeling" (to use C. S. Peirce's words) is pushed out, returning only as a dream of the mechanistic philosopher.[38]

If biology is to be made a branch of physics, it must at least be a broader physics that can explain biological phenomena. We have seen the outlines of such a physics: it does not simply break with classical mechanism, but also seeks to explain biological self-organization as part of a further elaboration of the thermodynamic definition of life (see chapter two). However, the self-reproductive aspect of life can also be approached from a more form-oriented, formal perspective of mathematics and logic. Can we use purely mathematical-logical considerations to discover the specific features of biological automata that make them self-reproducing?

Von Neumann's Automata

In the 1940s, the Hungarian-American mathematician John von Neumann (1903–1957) attempted to answer this question, and his answers helped qualify him as one of the fathers of artificial life. (Artificial life does not have very many mothers; perhaps women may not have quite the same aspirations since their relation to the creation of real life is better grounded in personally embodied experience). Von Neumann was not only known for his enormous productivity and his contributions to the mathematical foundations of quantum mechanics, game theory, and proof theory. He was active in military research during World War II, participating in the program to develop the atomic bomb. Finally, von Neumann contributed to developing the theoretical basis for the electronic digital computer and various computational methods. (Today's computers all have the sequential "von Neumann architecture.")

Von Neumann became interested in the general question of what kind of *logical organization* an automaton needed in order to be able to reproduce itself. The lectures he held in the late 1940s clearly show that he considered self-reproduction as the defining feature of living organisms.[39] Yet he attacked the problem less directly. Von Neumann had neither the possibility of nor interest in simulating a living system at the biochemical or genetic level. At that time virtually no one, including von Neumann, knew that DNA was the genetic material.[40] Rather, von Neumann hoped that he could abstract the logical form of the process from the natural material self-reproduction.

That he spoke of automata and not just machines requires some clarification. An automaton is a "self-moving" machine; a soda machine, for example, can via a single stimu-

FIGURE 3.1. Jacques de Vaucanson's mechanical duck (1735).

lus (a coin) autonomously execute its functions. Aside from various useful automata, history is rich with examples of humans attempting to animate automata, to fill them with life; in other words, to construct mechanisms with such a sophisticated repertoire of behavior that they resemble animals. Jacques de Vaucanson's mechanical duck from 1735, which could move its head, tail, and wings, as well as swallow "food," is a famous example.[41] The duck satisfied a Frankensteinian urge and fascinated people in the same way that the artificial nightingale in Hans Christian Andersen's fairy tale astounded the emperor and his court. Various control mechanisms were developed, such as the rotating cylinder with gudgeons used in music boxes to control the timely execution of behavior, and with this arose the possibility of *programming* automata to exhibit abilities such as writing or playing music.

These automata could not reproduce themselves, of course. However, together with the first adding machines, they were the forerunners of a generalized concept of automata or logical machines. An automaton could be considered functional according to the specific mechanical process it executed. This process could be specified by the abstract control structure or program that governed the automaton.[42]

Even though a concrete automaton might consist, for example, of wires, transistors, relays, and gears, the principles of its operation can be described in *formal* terms as a sequence of states and simple, step-wise movements chosen from a final register of possible states and steps. This specification is independent of what the actual components are made of. Viewed abstractly, the automaton is a set of physically unspecified states, input, output, and operational rules. By studying this set of states, we can determine what the automaton can do. In the 1930s, this was of special interest in the case of machines that manipulated symbols or numbers. Logicians and mathematicians such as Turing, Church, Gödel, and Post contributed to the path-breaking exploration of the mechanical computation processes. They became for computers what Watson and Crick became for biotechnology.

Their work led to a crucial new insight: that the essence of a mechanical process was not so much material-physical, as it was its program or control structure, which via an abstract set of rules—a formal specification—could capture the process's functionality. The machine has a logical form that can exist in many different manifestations.

It should be added that this movement led to the emergence of the generally programmable computer as the realization of a *universal machine*. The PC and other microcomputers that now exist in virtually all areas of society are examples of a machine type that in a certain sense is universal, because it is the input *program* that determines whether the machine will function as an advanced typewriter, a calculator, a bookkeeping system, a file cabinet, an illustrating machine, a game machine, or something else.[43] The very existence of the possibility of automated translation from one language to another also creates a demand for renewed efforts to construct formal theories about language, because procedures for translation can be fully formalized only when they can be handled by a program in a machine.

(However, it has become clear that it is precisely such features as everyday conversation, common sense, and creativity that have proven practically impossible to formalize satisfactorily.)

It was in this landscape of ideas that von Neumann developed a theory of automata and self-reproduction. His efforts can indeed be seen as an attempt to formalize the formation of life. Von Neumann's first bio-logical [*sic*] observation was that whereas the products of quite simple machines are often less complex than the machines themselves (for instance, a hammer is less complex than a machine that produces hand tools), the products of living organisms are just as complex as the "machine" producing them. It is even the case that during the course of evolution, increasingly complex beings were created. It must therefore be possible for nature's own machines to produce—over a sufficiently long interval—an "output" that is more complex than the constructor. In contrast, simple physical processes under a certain threshold of complexity have a tendency to decay to pure disorder (entropy grows in an isolated system, say physicists).

As a solution to the problem, in 1948 von Neumann presented an automaton that he believed could float around in a pond with free access to the necessary components, various "organs," which he described precisely. His student, Arthur Burks, called it the "kinematic model" because von Neumann excelled at descriptions of the components' kinetics or movement. Here we will present a short summary of the model, which expresses the logical conditions for self-reproduction. The self-reproducing automaton consists of components we shall call A, B, C, and D:

> A is a general constructor-automaton or "factory." It produces an output X when given an instruction or "logical description" $\beta(X)$ of the desired output. From this description, the factory chooses the suitable components in the pond of raw materials in which it moves around. Of course, the description must first be transmitted to A.

B is a general copying automaton or duplicator, which with a given description β as input delivers the description β, plus a copy, called β', as output.

C is a controller, which delivers to B an instruction, β(X), for double copying. C then appends the first copy to A in order to execute the construction of the described "product" X (here von Neumann conceived the construction as occurring during the destruction of the copy of the instruction). Finally, the controller attaches the remaining copy β(X) to A's output X, and then releases this (X + β(X)) from the machine A+B+C, retaining the original itself.

D is an instruction of a peculiar kind. D is a description that enables A to produce exactly A+B+C. In other words, D is the machine's self-description, D = β(A+B+C).

The automaton A+B+C+D, a synonym for the automaton A+B+C+β(A+B+C) thus produces an output that is precisely A+B+C+D (see the concept diagram, fig. 3.2). None of the individual parts are themselves self-reproducing, but each is necessary for self-reproduction. Thus, we see that self-reproduction is a system-property of the total ensemble of components, something that characterizes these elements' internal relation, their organization.

Even though the idea of formalizing biological processes

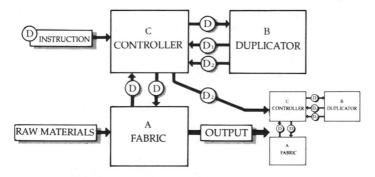

FIGURE 3.2. Conceptual diagram of von Neumann's 1948 model of automatic self-reproduction. The output (here shown reduced as identical with the "machine" A+B+C) is appended to the copy D_2 of the instruction.

is quite alien to biologists, it is fascinating to see how von Neumann came close to locating the inventory of the cell that years later would be revealed as the genetic code and the control mechanisms for the translation of genes into proteins. The cell's biochemical metabolism, which constantly produces new proteins, corresponds to A, the replication of DNA corresponds to B, the control of which genes must be turned on and off at different times corresponds to C, while the genetic information in the DNA itself corresponds to D in von Neumann's model.[44]

Von Neumann also considered the possibility of evolution via mutations in automata of this type. For example, were we to add an extra component called E, such that the automaton were now called A+B+C+D+E, and E were to be contained in the self-description $D = \beta(A+B+C+E)$, then a mutation in the description of E (changing it from E to F) would have no effect on the factory, the controller, or the duplicator functions. That is, it would not be a fatal mutation. It might instead result in another, perhaps better, automaton, A+B+C+D+F.

Now von Neumann was himself not entirely satisfied with the kinematic model. He was unable to adequately capture the minimal *logical* prerequisites for self-reproduction, for the model was still based on an intake of concrete, raw materials. This made it too difficult to elaborate a comprehensive set of precise, simple rules for the machine's movement. Such rules also had to include the ability of the automaton to recognize the correct components; in other words, it needed some kind of sensor mechanism, as well as a fine-tuned motor-control function which insured that they were put together in orderly fashion (these are extremely difficult problems that the discipline of robotics is now trying to solve). Von Neumann discussed the problem of minimal logical requirements with his friend, the mathematician Stanislaw L. Ulam, who proposed that another, cellular perspective might be better suited than the kinematic to logical-mathematical treatment of self-reproduction.

Von Neumann followed Ulam's advice and began to

work with what today are called cellular automata. Conway's Life game (p. 6) is one example of a cellular automaton: a "crystalline" lattice of cells or small boxes, where each box—in mathematical terms—constitutes a small automaton, that is, a little computer. With the help of a few simple rules, the box calculates its state in the next generation using the input obtained from the neighbor cells in the current generation. In itself, these cells have nothing to do with biological cells. They are mathematical objects, a visual representation of a certain kind of formalism. In their very structure, however, we can embed several interesting processes and forms (the glider presented in fig. 1.1. is one such example). All these processes and forms resemble, and perhaps totally correspond to, biological processes. Von Neumann sought to embed the logic of life in these abstract thought objects.

Even though formalism allowed von Neumann to bring the problem entirely out of the physical-mechanical domain and into the sphere of pure logic, the process was far from simple. In 1952–1953, he slaved over a large manuscript that contained a detailed draft of a solution. The manuscript consisted of a logical equivalent to the kinematic model with approximately similar "organs." These were located as a start-pattern in a massive matrix of cells, where each cell could have twenty-nine different states (rather than the two in Conway's Life game). The processes that in the kinematic model comprise physical movement are achieved in the cellular automaton (CA) model by transferring information from cell to cell (information is transferred when the cell computes its new state). Moving a structure in CA-space occurs by simply copying it in the direction of the desired location. The original is erased, and the procedure is repeated for a specified number of steps and follows all the automaton's internal rules (movement in a CA is thus "simulated").

The actual derivation of the solution is a multistage process requiring several complicated computer operations that need not be detailed here. In order to construct the actual rules (the table of state transitions) for the automaton that provided the desired behavior, von Neumann had to ensure that he allowed adequate computational and construction capacity in the cell-based formalism. There are two levels to the construction, and this gives some idea about the artificiality of this life: (a) the actual basic cellular array of automata, and (b) the "universal constructor" that is embedded in the basic automaton as a pattern of states. This pattern itself constitutes a virtual automaton, which in mathematical terms is based on a version of a universal Turing machine. This enables the construction of higher-order information processing, such as program-controlled construction, copying, transfer, and storage of information—all the features that were implicit in the kinematic analysis of the phenomenon of self-reproduction.

It is worth recalling that the information for the automaton's self-description (component D above) appears in two forms, precisely like DNA in a cell: as passive, uninterpreted data that are simply copied by the apparatus, and as active instructions that via interpretation help to control the construction of the machine's/cell's offspring.

Von Neumann never finished the manuscript, nor the design of the self-reproducing automaton. Arthur Burks filled in some of the gaps and showed that, in principle, the formal model could be simulated in a real computer.[45] It would perhaps cost too many programmers too much sweat, and, as far as we know, no one has actually succeeded in realizing this construction in any kind of real, physical computer. Von Neumann's model life has remained a torso. Edgar Codd and Chris Langton subsequently found more simple CA types, with far fewer states per cell, that can self-reproduce (see fig. 3.3).[46] Langton's construction is the simplest because he has designed it without having to embed any automaton of the universal-constructor type on which both von Neumann's and Codd's models were based (element A in fig. 3.2). It has the form of a loop that grows and divides, and whose offspring

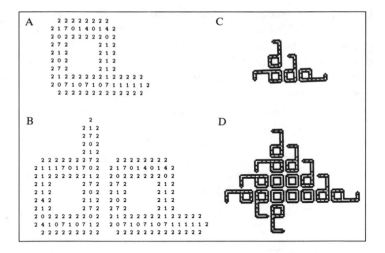

Fɪɢᴜʀᴇ 3.3. Langton's self-reproducing loop. A: in the upper left is the "seed," the cellular automaton's starting state. Each cell can have eight states (values 0–7). The actual state depends on the state of the cell and the states of the four neighboring cells during the preceding time interval, all according to the rules in a large table (not shown here). The seed has the form of a relatively stable sheet (the 2s) which encapsulates a data trace (the 1s). During the development of the automaton, a program runs into the track: the program is constructed out of two simple signals: the paired states 7–0 and 4–0. The signals are also transmitted into the "arm" to the right. For each 7–0 signal, the arm grows by one unit. For each 4-0-1-4-0 signal the arm turns toward the left. When a new loop has been created, the offspring are separated from the mother loop. The rules are so ingenious that the new loop repeats the cycle on itself. Below (B) is shown the configuration after the completion of 151 steps. The mother and daughter seeds continue their budding, and gradually create an entire colony, shown in C and D (from Langton 1984, see n. 46).

also grow and divide. The entire pattern in the CA-space of cells that change their states continually creates a veritable colony of individuals whose furthest member continues the growth process.

It is instructive to compare Langton's loop with a living cell. Here the genotype codes for the biochemical elements of a dynamic process that repeats the use of a set of rules. It is the process itself, also called metabolism, that computes the expression of the genotype, and thereby specifies the properties that come together to create the phenotype. One might say that Langton's model for self-reproduction corresponds to the fact that the information contained in DNA does not directly code for the phenotype, but for the many processes that together are responsible for the fertilized egg cell's subsequent development (or epigenesis) into an embryo and later into an adult individual. It is therefore tempting to conclude that Langton's model life indeed *realizes* life in the broader sense, even though the organism-environment relation is not integrated into the model.

Out with the Bath Water?

Many alifers have become especially excited about cellular automata. CAs exemplify beautifully the type of computation that is appreciated within artificial-life circles because it is thought that living beings process information in the same way as alife: parallel (many small units calculating simultaneously), bottom-up (see pp. 19–20), and with entirely local control over behavior. Together these principles allow for the emergence of system properties such as self-reproduction.

Von Neumann, in rethinking the self-reproduction problem logically, had anticipated the discovery of essential principles for the way living cells process information. These principles were made clear in detail in the years following his death in 1957, but the resolution was achieved in entirely different ways: namely, via molecular biology's combination of chemistry, biochemistry, and genetics. That von Neumann's anticipation of the principles of molecular genetics was possible at all immediately supports what Langton has called the principle assumption made in artificial life: namely, that the logical form of an organism can be separated from its material basis of construction, that its "aliveness" is a property of the form, not of the materials.[47] A truly "form-al" approach to life, we might say.

Not unexpectedly, this postulate of separateness (or medium-independence) seems quite alien to molecular biology, for molecular biologists are preoccupied with the connection between the material, chemical-molecular foundation of life and its higher level informational, genetic, and morphogenetic processes. An orthodox molecular biologist would deny that the tradition of von Neumann to Langton adds any new knowledge to biology at all. Moreover, he would believe that biology must always be based on fundamental knowledge about the actual chemistry of life. He would argue that it is biochemistry and organic chemistry that connect biology to the other natural sciences. There are profound philosophical disagreements expressed in the discussion between alifers like Langton and biologists of the traditional materialist conviction. The latter are skeptical about a total redefinition of the very object of general biology when it is not just about life as it actually is, but life "as it could be."

It will be interesting to see how this challenge to biology's philosophy and theory will be met. Biology has been characterized by reductive views that identify living beings entirely with their biochemical content, thereby complicating efforts to understand organisms, species, and ecosystems as complex wholes. A synthetic understanding like that of Darwin or the Odum brothers (in ecology) has not exactly dominated twentieth-century biological research. Biology thus has been vulnerable to a holistically oriented criticism from outsiders, often philosophers, who have found it easy to point out biologists' failure to understand the dynamics of coherent natural systems. Now, after the philosophers, come the alifers and the computer scientists, who offer neither romantic syntheses of a natural philosophical kind nor ecological holistic pictures but instead synthetic life. This possibility will perhaps generally be ignored or rejected as too outrageous.

Yet is it possible to completely abstract form from its material content? Von Neumann's remarks on his own formalization of the problem of self-reproduction are illuminating here. He believed that by formalizing the problem in this way (in the CA model), "one has thrown half of the problem out the window and it may be the more important half."[48] This means that we renounce trying to explain how the parts are in fact put together out of "real things," how they are constructed out of "actual elementary particles," as he expressed it, or higher chemical, biochemical, and other material units. In contrast, Langton does not believe that the baby has been thrown out with the bath water: one has to distinguish between how life as we know it emerges from physics and chemistry, and how possible life and life-like behavior, can emerge from local interactions at a lower level in a population of logical "primitives," the units used by logical analysis (an automaton, for instance).

The question remains as to whether the logic of life that we derive using abstraction and logical analysis has any relevance in the material world. Langton himself points to the problem that Codd's and von Neumann's models of self-reproduction require universal construction.[49] Universal-construction capacity for an automaton is an effective means of avoiding entirely trivial forms of self-reproduction, e.g., very simple patterns that generate themselves by a domino-like effect along the lattice of cells. Little is gained by formalizing these patterns. However, the universal-constructor criterion also rules out naturally occurring self-reproducing systems. Animals and plants cannot be made to yield any desired output just by encoding it as a description in their DNA. The criterion does not have much to do with biology; it is a biology of the impossible and is ultimately fatal for any assertion that von Neumann's model describes something other than a very abstract logic of self-reproduction. The abstraction loses something essential.

One way of solving the problem is the apparent solution that both the material and the more formal, logical elements must necessarily enter into a general *bio-logical* description of life processes. Life is not only digital. The material aspect must be included because neither life nor logic arise out of nothing. Life requires a material foundation and a historical process that organizes the material on higher levels and that places limitations on the logically possible ways in which life can behave. It is possible that evolutionary logic, in the form of Darwinian natural selection, may be valid not just for earthly life-forms but for those on other planets as well. This evolutionary logic is closely conditioned, on earth at least, by a rich biochemical substratum for the genetic coding of the genotypes selected, a substratum comprising the material basis for growth and development on Gaia, the living planet.

WITH LIFE AT STAKE

One can speculate why digital life is so fascinating. If the dream of life's creation lies within the cellular automaton, what is this dream made of?

Perhaps it arises from a Pygmalion desire. An ancient Greek legend tells of a Cypriot king named Pygmalion. He fell in love with a statue of a beautiful woman—perhaps the goddess of love herself—whom he himself had sculpted. Pygmalion became so enamored of his work that he embraced it. He begged and pleaded with Aphrodite for a wife of the same appearance, and at last she took pity on him. She answered his prayer by making the statue come to life.

As with stories from real life, the myth explains nothing but says much about efforts to create life artificially. When Conway introduced his Life game, it quickly became extremely popular among computer enthusiasts. And for good reason: the game is quite simple to understand, easy to program, fun to play, and quickly generates a feeling of being a cocreator of the universe of possibilities that unfolds in such a cellular automaton. You can play God in your own universe.

Conway himself created the two rules of the game (see chapter 1) based on considerations of what could produce interesting structures. Through a series of experiments with variations, he investigated whether three requirements for a set of rules could be fulfilled simultaneously:

1. There should be no starting patterns that lead the population of turned on cells to simply grow without limits, explosively and unrestrictedly.
2. Nor should there be any start patterns that even seem to grow in such a way.
3. Instead, there should be start patterns that grow and change at a suitable interval before eventually finishing.

The end could occur due to their disappearance (caused by overpopulation or isolation); or by their falling into a stationary pattern without any more changes (a quiet life); or by their entering into a cycle of state transitions in which the pattern fluctuates between different states over periods of two or more generations. An example of the last is the "blinker," of three horizontally aligned turned-on cells that after a generation with three turned on and vertically aligned, returns to its original, horizontal arrangement (i.e., the period is 2).

The "glider" is an example of a structure that is both periodic (with the period 4; see fig. 1.1) and progressive. Since the "velocity" in which a king is moved in a chess game—one square per move—corresponds to the most rapid movement in the Life game, Conway called it "the speed of light." The glider moves along with a quarter of the speed of light; no pattern in the game can copy itself forward at the full speed of light.

The reactions Conway simulated using the dishes on his own checkered floor quickly stopped. Most of the starting patterns pulsate for a time and then die out or settle into cycles of periodic repetition. There was no continuing growth. When the Life game was first presented in *Scientific American*, in October 1970, Conway therefore guessed that a population could not grow without limits (that is, a starting pattern with a finite number of cells could not surpass a finite upper limit for the number of turned-on cells).[50] Yet he was unable to demonstrate his assertion, and offered a prize of fifty dollars to the first person who could prove or disprove it by the end of the year. Conway was aware that one possible way to disprove his postulate would be to discover a "glider gun," i.e., a configuration that constantly fires gliders or other moving objects out into the space of empty cells, or a structure analogous to a "puffer train" that leaves a trail of "smoke" while it crawls forward. The investigation of these and other strange life-forms thus commenced.

One of the programmers who participated in the search from the very start, Bill Gosper, today says that he still uses his spare time to explore the Life game's universe. It is a universe that contains innumerable surprises. Gosper remarks that, "at first it was not clear that the things that could happen in this universe were nearly as complicated as the things that can happen in our universe. And subsequently, just by a sequence of small discoveries, just by degrees, it became clear that anything that we can describe can happen in the Life world."[51]

Gosper won Conway's reward for his sought-after life-form in November 1970. Gosper's research group for artificial intelligence at the Massachusetts Institute of Technology composed a program that showed the updated states of each generation on an oscilloscope.[52] It was here that Gosper discovered a glider gun. Figure 3.4 shows the gun in the process of shooting out gliders, one new glider for every thirty moves. Given enough time (and a large CA lattice), this pattern will evolve into infinitely many on cells. Gosper's group also found another structure, a "pentadecathlon," that could either eat or return gliders according to how it encountered them.

It soon became apparent that Gosper and several other Life game enthusiasts were on a veritable journey of exploration into a strange new world. Conway's proposed "puffer train" was also found, both in an environmentally safe edition, where the smoke trail was quite small and regular, and in a dirty, spouting version where the train left a greasy trail of fuming smoke.[53] The start structure of the

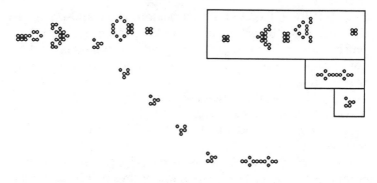

Figure 3.4. The large box at the upper right depicts the glider gun in its initial state; in the middle box is the "pentadecathlon," located so that it "turns" the gliders 180 degrees; the little box shows a glider. On the figure itself is a snapshot of the pulsing gun after 137 moves, where it has produced nearly five gliders. The gun changes form periodically while producing gliders. The first glider soon meets its fate when it is turned 180 degrees by the pentadecathlon, and on its way back toward the cannon it is destroyed in a collision with the next glider. Only the "on" cells are shown, as circles.

dirty train, suitably enlarged in fig. 3.5, can be seen so that we can check the rules for a selected cell in the first steps. Figure 3.6 shows the pollution trail itself after a nice long journey, suitably reduced.

Also discovered were several versions of Life in three dimensions.[54] These appear clumsier and are difficult to grasp, since the objects' front surfaces mask what is happening in the back. (This disadvantage is shared by our own three-dimensional world: we cannot, fortunately, see everything in a single view.) Moreover, in purely mathematical terms, there is no real limit to the number of dimensions of Life that can be adopted; however, it becomes simply impossible to "see" and much slower for the computer to compute. Hence, multidimensional Life does not have much attraction. The "normal" two-dimensional Life is enough to allow for the discovery of undreamed of complexities. When Gosper states that anything can happen in this simple cellular automaton universe, he means it.

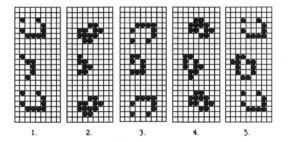

FIGURE 3.5. The first four steps of a "dirty puffer train." It consists of two "spaceships" (the two prostrate Cs above and below) plus a configuration in the middle that, interacting with the spaceships, leaves a long trail of smoke (see fig. 3.6). If the middle configuration of the dirty puffer train was allowed to evolve for 150 generations without being limited by the spaceships, it would have become the following structures: some "quiet life" consisting of three blocks (of four on, appearing as ::); a "ship" (of six on cells that lay diagonally), both crystalline; plus two gliders, each marching to their respective corners of the world.

FIGURE 3.6. A "dirty puffer train" after four hundred steps, reduced as compared to fig. 3.5. To the extreme right one can see "the front" (fig. 3.5) move onward toward the right, leaving in its path a trail of "smoke," which after a period with whirling movements falls into a calm, static pattern (only the on cells are shown, as dots).

Has anyone in a Life game actually seen any of the genuine self-reproduction as formalized by von Neumann's model? ("Genuine" as opposed to the simply trivial, periodic replication of a simple structure, as in the glider's diagonal movement.) Not yet, but only, Gosper believes, because today's computers do not have enough computational capacity. What we see here corresponds almost to the level of nuclear physics.[55] However, if we have the real "atoms," we can in principle also build an organism from them. The same is true of the Life game. In fact, it can be shown explicitly that it is theoretically possible to implement von Neumann's complicated self-reproduction in this simple game.

This can be done by taking a detour. The detour is a simulated computer. It can be shown, in fact, that the Life game can be used to construct a full-blown, virtual computer, a "genuine calculator," which processes information on the cellular fields but which is simulated within the Life game universe. The trick consists of using Gosper's glider gun and several other elements as basic components. The gliders, for instance, are used as the units that transmit and store information. (One can just think of several gliders—as in fig. 3.4—and their greater or lesser space as a Morse code.) Computers have been constructed with different material components without altering their fundamental principles of computation. John Conway has shown that logical gates like AND, OR, and NOT, which are necessary for a computer's function, can be constructed with just a bit of skill from some basic patterns in Life.[56] Two gliders that are at an angle to each other and collide, in so doing dissolve each other. The NOT gate can then be constructed from a glider cannon that fires gliders at a right angle towards an input consisting of an information-carrying glider stream whose individual gliders and empty space (e.g., "GG GG GGG G") represent, respectively, zeros and ones (i.e., "11011011101"). When the cannon's gliders crash perpendicularly into a glider (a "1") from the input, both dissolve into nothing, corresponding to the creation of a space (a "0"). When the glider encounters a space ("0") from the input stream, it simply continues onward (corresponding to a "1"). The input "11011011101" to the NOT gate therefore yields as output "00100100010" (see fig. 3.7).

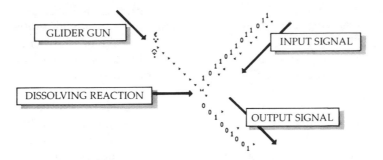

FIGURE 3.7. A NOT gate constructed in the Life game. An upright glider gun at the upper left (compare fig. 3.4.) fires gliders toward an input signal of gliders (a periodic 110110110 ... where "0" is denoted by an empty place in the glider series). The signal is negated by the output (001001001 ...).

Even though it will take a long time, these kinds of exercises are enough to show that a universal Turing machine can be built within the framework of the Life game. The two rules (see p. 6) are the basic ones, but by using the various Life game patterns such as the glider gun, one can implement a Turing machine at higher levels. This Turing machine can then itself implement anything that is at all computable (which includes any of the programs on today's ordinary computers). Other cellular automata such as von Neumann's, Codd's, and Langton's self-reproducing machines are precisely computable. This Turing machine is a kind of computational Chinese-box system. It would probably demand a Life universe larger than any computer can deal with at present. Von Neumann calculated that the total size of his self-reproducing CA automaton, if it were to be realized, would be two hundred thousand cells. William Poundstone has calculated further and found that constructing the CA automaton in the Life game would require on the order of 10^{13} cells, or pixels, if we let a pixel (one dot on the computer screen) correspond to a cell in Life. If we make one pixel equivalent to one square millimeter, the screen would be three kilometers long, with an area of about twelve square kilometers, one fifth of the entire area of Manhattan! It would be impossible to handle these Life-generating processes in any detail. It would look like a thick fog. But the patterns created here would be reproduced and grow independently until someone turned off the electricity.

We see here a dizzying structure of levels on top of levels of computations, which (in principle) make life possible, and which make for happy days in a mathematical world. It is possible that we still can find new shortcuts to make self-reproduction in Life easier, thereby avoiding complicated Turing machines or universal constructors or Langton loops. The next chapter examines the diversity of computational tools that a-life research has brought to theoretical biology.

Chapter Four

ARTIFICIAL GROWTH AND EVOLUTION

THE DEVELOPMENT of powerful electronic computers has pushed mathematics into a new golden age. We have witnessed the birth of new geometric objects with strange names like Mandelbrot and Feigenbaum as well as other fractals with exquisite forms. We have also been presented with chaotic systems, strange attractors, cellular automata, and artificial life. A wealth of bewildering structures has been discovered, structures that had been impossible to imagine just twenty years ago without the computational power and image-making capability of the computer. Complicated solutions for systems of coupled differential equations can be visualized in the same second as the machines calculate them. We can take a journey of discovery into a mathematical landscape of patterns and forms that, once the rules and initial conditions are defined, unfold themselves on the screen that masks the number-crunching microprocessors. This development makes mathematics into something other than proofs of formulas and derivations of theorems. Mathematics comes to resemble more an experimental science (to the regret of the strict formalists among mathematicians).[57] The computer has become an observational tool in this process. We can observe entirely new universes of forms, including "life-forms" of synthetic biopatterns that behave like ants, plants, cells, or other forms of life.

The scientists working with cellular automata assert that they feel like Anton van Leeuwenhoek when he constructed his first homemade microscope and gazed down into a drop of water to discover a world of fantastic beings. During the late 1600s, van Leeuwenhoek investigated water, earth, slime, sperm, hair, and the body's internal organs, all of which hid unique microcosms. When the Dutch optician and town-hall caretaker discovered microorganisms inside droplets of water, no one believed him. Similar disbelief greeted the renowned natural scientist Robert Hooke when, in a quite angry meeting of the Royal Society in London in 1677, he used a similar microscope to show tiny crawling creatures in a glass of fermented pepper water. The members were somewhat shocked. The microscope opened up a previously unknown reality. Throughout the scientific revolution, the microscope was as important for the spread of a new worldview as the telescope was for astronomy.

Today it is the computer, an instrument that permits us to see the deepest structures, that has taken the microscope's place as the predominant *speculum mundi* or mirror of the world. It is, of course, a world based on mathematical formalisms, but formalisms that nonetheless contain such dynamic forms of movement that they hold the promise of being able to describe both the quivering of earthly life and the complex mechanics of the heavenly bodies. As in von Leeuwenhoek's time, there exists today a degree of skepticism about the kind of images of the world these new instruments are revealing to us. The instruments are not unreliable, nor are the images they create simply wrong, but there exists considerable uncertainty as to what kind of world they are actually describing. Is what we see simply a human construction, or do the computer-generated pictures describe an independent reality? We may again pose the age-old question of the relationship between mathematics and reality. Alifers respond: the pictures that computers construct are the world of the biology of the possible, a synthetic world of artificial life that, despite its artificiality, must conform to the universal laws of self-organization and evolution. It is precisely thus that the study of artificial life can help provide us with a more complete picture of the general laws for complex systems. Such laws of form are still covered by a veil. Biologists—whether wearing

boots out in swamps or using microscopes in the laboratory—have only lifted a tiny corner of this veil. We shall see
some remarkable examples of artificial form-generating
systems.

THE ALGORITHMIC BEAUTY OF PLANTS

Goethe and the old Romantics knew that it is the laws of
form alone that govern the growth of plants and unfolding
buds into complete leaves with species-specific patterns. In
our century, the biologist and classicist D'Arcy Wentworth
Thompson attempted to combine the Romantics' intuition
with a mathematical description of the underlying architectonic principles for the growth of animals and plants. However, when Thompson was elaborating his major treatise on
the subject, *On Growth and Form*, biology was still in its in-
fancy and computer science nonexistent. Hence, Thompson
had to restrict himself to simpler, static patterns: he formulated equations that described, for example, the corkscrew
lines of the cauliflower head, the beautifully curved shells
of snails and mussels, and the bridge-like constructions of
bone tissue.[58] The transformations of the forms he studied
were geometric depictions of already-created forms among
various species of fish and mammals.

The situation was different for the Dutch biologist Aristid
Lindenmayer. Like Thompson, Lindenmayer was interested in a formal, mathematical description of plant growth.
Working some decades after Thompson, Lindenmayer had
the benefit of being able to utilize both the new knowledge
of genes' function (the "genetic program") as well as the
computer's ability to handle formal grammars that had
been developed to study the structure of language. In the
late 1960s, Lindenmayer developed a formalism now
known as L-systems. While the so-called Chomsky grammars for language could be used to simulate the creation of
long, embedded expressions in a formal language (as when
a sentence, e.g., a parenthesis like this one, is made longer
[or even longer]), Lindenmayer succeeded in creating a

system that could not only generate longer one-dimensional sequences, but that could apply its rules alongside each other in several places at once. Whereas a text is a linear sequence of words, organisms evolve in parallel fashion: many cells in the body are created simultaneously; the tree creates its branches in several places at once. Cellular automata also execute parallel computation, but the patterns they create typically extend themselves only to the most forward row of cells. Trees and bushes create new branch shoots and expand in all directions at once.

They do this in very different ways. The worlds' plants have an infinite number of ways in which they grow. Nevertheless, there are certain constant relations that define a plant's topology: every plant has a basis from which roots emanate downward and a central axis grows upwards (the stem of a tree or the stalk of an herb). On the main axis (the axis of zero order), there may sit on certain branching points leaves or axes of the first order, and on these axes of the second order, and so on. Every axis has a base and a top. Leaves are always terminal: new axes or stems never sit on the edge of a leaf or on its top. These uniform conditions constitute the basic relations in the formal system that Lindenmayer developed in order to describe the rules for new growth of the individual elements. This system can be viewed as a series of recipes or algorithms for the creation of the parts of a plant. An example is the little tree in fig. 4.1a. The tree is represented in fig. 4.1b as a series of oriented axes and can in turn be written in one of the formal L-systems as a string formula where the brackets [ ] define the branching-off points.[59] Hence: A [B] [C [D] E] F [G] [H].

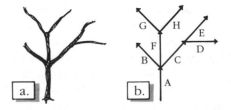

FIGURE 4.1. Formalization of a tree.

Why make the branch structure into a formula? L-systems rely on a principle that we can call "rewriting." Rewriting is a technique whereby complex objects are defined by replacing parts of a simple start object with another object according to a specific rule, and then repeating the procedure for the object that appears. For example, the ice-crystal–like curve called Koch's curve is constructed from an equilateral triangle that is the start object, and a generator consisting of a broken jagged line: _⋀_ (of four small, equal fragments), and a rule for rewriting these that replaces each line fragment —— in the object (each side of the triangle). Each time the rules are applied again (that is, recursively) the new object attains four times as many line fragments. The same principle is used in recursive programming.[60] The jagged sections of the curve become steadily greater in number and finer in scale, so that one soon has to have a magnifying glass in order to see the details. When enlarged, they resemble themselves at all levels. They are *self-similar*.

Many plants have forms that are strongly self-similar, as in the branching of an evergreen tree or the leaves of a fern. When a section of the tree or fern is enlarged, one finds the same pattern repeated over and over. This can be used when the L-system has to specify a plant by rewriting. A simple piece of the axis is replaced by a piece of three segments having two side branches (see fig. 4.2), and this process can after a suitable period of growth be repeated again and again.

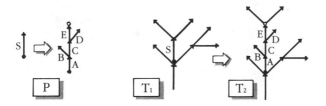

FIGURE 4.2. A rewriting rule p and its application on the piece of axis S in the tree T1 (from Prusinkiewicz and Lindenmayer 1990, see n. 59). The same rule could now be applied recursively on the new axes A, B, etc.

When the angles are permitted to vary, this technique will produce an infinite number of highly realistic branching forms that can be generated on a computer (fig. 4.3). Together with the computer graphics expert Przemyslaw Prusinkiewicz, Lindenmayer created a paradise of a flower garden on his computer, which in a hyperrealistic way draws beautiful lilacs, carnations, sunflowers, and several other kinds of ornamental plants. It is not surprising that there is something cartoon-like about these exquisitely perfect pictures. They seem simply too perfect and unblemished, while the black and white computer-generated sketches appear more honest. However, it is not much more than a question of time and computational efficiency before more "authentic" plants will appear. Prusinkiewicz and Lindenmayer's colleagues have experimented by allowing the plants to appear more natural by making them less perfect and somewhat asymmetrical: the computer can be instructed to "draw lots" about what kind of branch or leaf will be depicted in the next computational step of growth (a random-number generating algorithm decides the unfolding of the individual parts of the algorithmic plant).

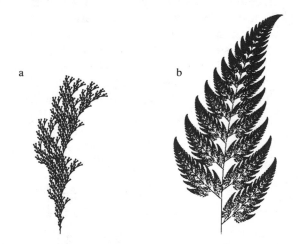

a   b

FIGURE 4.3. (a) Examples of plant-like structures drawn on a computer using one of Lindenmayer's formal rules (from Prusinkiewicz and Lindenmayer 1990, see n. 59); (b) Other techniques besides Lindenmayer systems may generate self-similar structures; shown here is a computer-drawn fractal fern leaf from M. Barnsley's "chaos game."

Do the swirls from computer graphics capture the plants' real dynamics? Certain common features exist at the general level: both the structures created by natural plants and the L-system's artificial branches, leaves, and flowers evolve in parallel, following a true locally operating logic. Both cases involve the application of simple rules, genetic and algorithmic; the rules are applied over and over to produce an often unpredictable beauty in what might be called patterns of great logical depth.[61] The computer plants often resemble their original forebears quite accurately. Apparently, the scientific mimesis within this entirely unique and realistic genre is quite successful.

If nothing else, the technique can be used in animated films that use realistic imitations of natural scenery. It is a seductive artificial reality that is not art, but that could be the medium for possible art, though art will hardly improve by a total identity with its object. This is reminiscent of the story of the man who asks Picasso, "Why don't you paint people the way they really look like?"

"What do you mean?" asks Picasso.

"The way my wife looks on this picture, for example." The man takes out a photograph.

Picasso replies, "She's rather small, you know. And quite flat!"

As the alifer Peter Oppenheimer remarks, the fact that a computer plant resembles its original forebear is far from a guarantee that the algorithm that has generated the image corresponds to the mechanism that has generated the plant.[62] That is, the image may look right for the wrong reasons. Fractal images of mountains that are created via recursive subdivisions of originally smooth or regularly edged surfaces may appear quite convincing, but we know full well that mountains' rugged surfaces are not created in this way. The images reveal more about how we sense mountains than they express any theoretical truth about the mountains' geology. Our obsession with computer-generated plants is just as much a matter of the observer as of what is observed. There is no doubt that a plant's apparently complex structure can be the result of a repeated (recursive) application of a relatively small number of genetic rules. The Austrian anatomist Rupert Riedl points out the necessity of economizing information in the coding of an organism: if the code for the creation of a lobe of a fern leaf or a single pine needle can in any way be said to be (indirectly) represented in DNA—that is, written in the form of an epigenetic code corresponding to an algorithm for a pine needle—it is fully sufficient that this code exist in one single place in the entire genome (the total genetic material found in every single cell in the pine tree).[63] This "pine-needle algorithm" can then be applied by the tree again and again, each time a pine needle is to be created. Used here, "algorithm" is simply a metaphor, for in real life the creation of form is not necessarily specified in the same explicit way as a computer's algorithm specifies the form of an artificial flower.

Moreover, we must not forget the level or aspect of the biological system we wish to depict. Are we dealing with the creation of the general form of the branch structure in a pine tree, the distribution of needles on a new branch, or the individual pine needle's own inner structure and external form? Lindenmayer's systems often work with the macro structure, the whole. A pine needle, however, is itself a tiny, complicated whole of component parts. It requires other models.

The Riddle of Form Generation

If a pine tree must utilize the principle of recursive programming to create many identical needles, what about an animal? How does nature create the specific form of a dog, a maple tree, an antelope, an eagle, a centipede, or a human? Recursiveness alone cannot explain how the exact form of a finger, a pine needle, a maple's leaf, or a centipede's leg is created during the transformation from fertilized egg cell to adult organism. Recursiveness only tells us that the same routine must be called up and executed several times. The routine itself, understood as a rule in a certain part of the epigenetic code, has remained unknown.[64]

This is the essence of biology's riddle of morphogenesis, the generation of form.

We do know that the forms generated are species-specific—no human is born with chimpanzee-formed bones—and hence, also inherited, i.e., imprinted in DNA. What is lacking for humans, however, is this simple description of DNA in detail. It is a sizable, but nevertheless feasible task, which at the moment is being pursued by the international Human Genome Project, in which molecular biologists throughout the world are participating. (Our total genetic material is estimated to contain about three billion bases or "letters"). It is tempting to believe that because we can "read the machine code" for the human organism, we can also understand its architectonic principles. But this is not so easy.

The problem is that we have been induced into viewing an organism as a bundle of properties that are inheritable. From genetics, however, we know that an inheritable trait—the ability of the hemoglobin protein to bind oxygen in the blood, for example—derives from a specific gene. The problem here is that "traits" or "properties" have many meanings, only some of which denote properties that are inherited. Because several biochemical properties are effects of specific proteins (and therefore encoded into one or a few genes that can be localized to specific points within the genome's DNA), it does not follow that all properties have this particular character, which is derivable from the functioning of one kind of protein molecule.

The form of a hand or the structure of an eye are examples of extremely complex "traits" that could be selected or distinguished by the observer in other ways (where does the "hand" end and "wrist" or "arm" begin?). Furthermore, a complex trait has several requirements: various proteins, cell types, tissue types, their three-dimensional structure, and the entire process that has generated the structure. The problem of form concerns the differentiation of cells and the regulation of this process in time and space. It is not enough that a fertilized egg cell divides, forms an embryo, and that the cells differentiate themselves into approximately 210 different cell types. They must also be placed correctly. Your foot and hand contain exactly the same cell and tissue types but, fortunately, each has a unique form.

In more abstract terms, the problem of form also concerns understanding the development of complex, dynamic systems. It is characteristic of these systems that their traits can be *coded in the whole*. The properties arise within a complex system where many simpler units interact, often with regulating and self-reinforcing feedback linkages, and the very conditions for the holistic properties are themselves produced by the system's concrete history. This means that the emergent properties, instead of being represented in any central master code, are constructed anew each time an organism is created.[65]

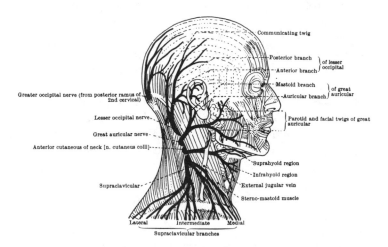

FIGURE 4.4. Anatomy of the human neck and lower face. Figures like these illustrate the problem of biological form generation to its full extent. The principles for three-dimensional structuring of the anatomical jumble of nerves, blood vessels, muscles, bones, joints, tendons, skin, and of the tissues these consist of have not been found yet. The sequence of the building blocks that make up the proteins is coded explicitly in DNA. However, the DNA contains no direct information about how these proteins then create a whole having a characteristic form (muscle cells, nerve cells, etc.) and how these cells are organized into tissues. (From V. Pauchet and S. Dupres 1976: *Pocket atlas of anatomy*, Oxford: Oxford University Press.)

It goes without saying that when two rabbits mate the result is a rabbit, but this is not because rabbitness is explicitly imprinted into DNA. Rather, it is because the same species characteristics are constructed anew each time a rabbit-specific egg and its rabbit-specific proteins and DNA define the initial conditions for the process, where the fetus slowly produces its specific form. It is not written in any particular place in the DNA molecule that a rabbit has two ears and is precisely this length and shape. The new construction of a complex system from identical rules and identical starting conditions can produce a somewhat identical result without explicitly coding the macroscopic construction. What is explicit about the coding lies at the microlevel, in the primary structure of the individual proteins.

It is thus hinted that the problem of form in biology cannot be solved using the methods of molecular biology alone, which tends to focus on processes inside the individual cell, but must instead be supplemented by other perspectives. One such perspective involves the study of the types of complex, dynamic systems that create special patterns. The Life game and cellular automata generally comprise such examples.

Pattern Machines and Models

Stephen Wolfram at the Institute for Advanced Study in Princeton is one of those who has continued to work with von Neumann's theories about cellular automata and has created an enormous interest in their ability to function as models for complex natural systems. When Wolfram gives a lecture, he often takes with him a few conch or mussel shells in order to demonstrate that the patterns he can obtain with automata are nothing more than nature's own patterns.

The cellular automata that Wolfram has developed are often one-dimensional, but visually "extended out into time." Conway's Life game is a two-dimensional automaton, and running it on a computer typically depicts the momentary state of all cells in the whole plane.[66] Wolfram's automata consist of an individual row of cells, but his computer visualizes their temporal development such that the states for the next generation are written below the preceding state, shown in different colors for each kind of state. This means that as the development of the cellular automaton proceeds, it drags with it a tail of cells (their previous states), so that it appears as if it were growing, line by line down the screen, much like the patterns of ice "flowers" on a window slowly creeping down the pane. Figure 4.5 shows a few examples of Wolfram's automata, some of which quickly die out, while others create very complex patterns.

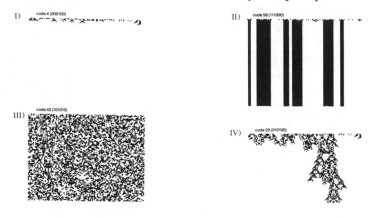

FIGURE 4.5. Four qualitatively different types of one-dimensional cellular automata can be observed: automata which are (1) rapidly dying out in a monotonous state (white all over) [code 4]; (2) creating stable, highly ordered, crystalline patterns [code 56]; (3) chaotic patterns [code 42]; and (4) complex localized structures that "live" for a short period of time and then die [code 20]. (Taken from Wolfram 1984c, see n. 71.)

Seen as dynamic systems, cellular automata can be used as a conceptual model to show general features of morphogenesis (biological-form creation).[67] In this way they supplement the older models and concepts of form generation elaborated by biologists such as C. H. Waddington. This application is thought provoking, because it shows how more intuitive metaphors and abstract notions in science, as we shall see in the case of Waddington, can be replaced with more precise mathematical models, even though these, too, may be quite metaphorical, and thus have no guarantee of being more valid models for the generation of form.

Waddington, an embryologist, once said that the development of the embryo could be understood by analogy with the unfolding of the entire spectrum of Euclidean geometry from a few simple axioms and rules for the derivation of geometric proofs.[68] He used this image to emphasize what is constructive about the creation of form. Criticizing the preformationist view that embryo development simply consists of translating a fixed, ready-made blueprint in DNA, a pattern that already contains the form, Waddington also invented another strong metaphor: the epigenetic landscape, a visual model of the dynamics of embryonic development (fig. 4.6). The model imagines a ball that represents the embryo rolling down through a curved landscape with hills and deep, separated valleys; these correspond to different developmental paths for the embryo and its cells.

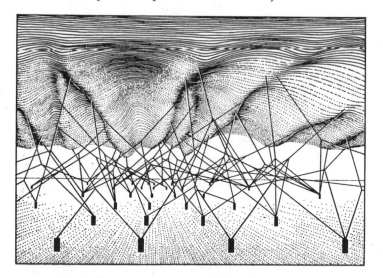

FIGURE 4.6. The epigenetic landscape. An illustration of the gene's modulation of the landscape's form. (After Waddington 1957, see n. 65.)

Under the terrain, Waddington sometimes depicted a network of genes that together determine the height of each of the hills and can thus influence the ease whereby the ball, subjected to minor, random disturbances, can be pushed over into a new pathway on its way through.

Waddington's epigenetic landscape emphasized the necessity of seeing the effect of individual genes in association with the entire genetic substratum of interacting genes ("the genetic background") and the additional conditions defined by the complex developmental system. Furthermore, it called attention to the random developmental noise or more violent environmental disturbances that could push equilibrium over into another path, resulting in a very different final product (the ball rolls down to a completely different place), even with an identical set of genes.

Waddington himself made an attempt whereby ether treatment of fruit-fly larvae could induce the development of adult phenotypes that were exact copies of the kinds of bodily defective flies otherwise thought to develop only by mutations (so-called "phenocopies"). He further demonstrated that by continually changing environmental conditions, he could cause the genetic assimilation of these altered phenotypes; that is, the change would become stabilized via subsequent genetic changes. This was not accomplished by any Lamarckian inheritance but by a form of feedback between selection in the changed environment and embryonic development via other routes in the epigenetic landscape, which geneticists had not previously been aware of.

The epigenetic landscape operates as a very flexible and general conceptual model, helping to systematize the interpretation of Waddington's many experiments and theoretical speculations. Its limitation is that it says nothing at all about how the genetic substrate could form the landscape itself, only that it must somehow be able to do so.

Here the cellular automata models are much more precise, usable, and well-defined, mathematically at least. We have seen that any specific state in the automaton's cells is totally determined by the states of the neighboring cells in the previous generation, and by a detailed rule or "state transition function"; i.e., a long table that prescribes that "if neighboring cell x = red, y = blue, . . . q = black, then the next cell state = yellow; if neighboring cell x = blue . . ." and so on. The point of departure is this set of rules together with an input, a given initial state of the first row of cells. The input is a form of spatial, structural information. The set of rules, using an analogy of a fertilized egg cell, can be termed the automaton's "genome," and the DNA instructions would be the "program." The initial state of the automaton ("What colors are the first line of cells?") is analogous to the biological cells' initial state with respect to cytoplasm, ribosomes, and the entire machinery or set of data upon which the DNA program must now operate: the epigenetic substrate.[69] In egg cells of many species (amphibians, for instance) it has been shown that the spatial organization of the individual cell components themselves represents an important piece of spatial information that determines the orientation of the coming embryo's different symmetry axes (up/down, right/left, front/back). As in the L-systems, the same rules are used over and over for each updating of the status of the automaton cells, and a pattern quickly emerges in the course of development.

We can also consider the automaton's development as a form of computation. The sequence of initial values in the automaton is interpreted as data, just as the sequence of binary digits in the computer's memory. During the automaton's development, information is processed; the states of the automaton cells are modified according to the rules, corresponding to the rules built into the computer's central computational unit, its CPU. The calculation, the automaton's evolution, is irreversible. Hence, the trajectories that describe the automaton's possible development in the total space of states will run together and be pulled in the direction of certain attractors that constitute a relatively small part of the total universe of possible states. An attractor can show itself on the screen as a pattern. It may be a form with a certain degree of stability, either boring and monotonous or radiantly beautiful. The cellular automata are pure pattern machines.

Regardless of what random initial condition is chosen, the system will organize itself in the direction of these attractors. Wolfram has played at being a computational zoologist and has classified different kinds of automaton rules according to the kind of behavior they produce. Automata may evolve into four distinct types of movement:

1. monotonous, boring, "dead" states
2. simple, stable, or periodic structures
3. chaotic patterns where the automaton is attracted by so many different stable states at one time that it is very unstable
4. complex, localized structures that evolve dynamically over a certain period, until they perhaps suddenly cease to exist as patterns.[70]

Some computational tasks are simple: they will quickly stop, corresponding to one of the first, trivial forms of automata. Other computations are more complicated, corresponding to a long, complex, behavior of the cellular automaton. Even though one can always determine the result of a given number of steps in the development of a cellular automaton—by explicitly simulating the development (executing the computation) for each step—there does not always exist a more effective procedure that, as it were, would make a shortcut. If such a shortcut does exist in the form of a mathematical formula, it can be used in order to discover whether the automaton is still alive in, for example, the 6,294,578th generation. One would not need to simulate over six million steps before finding the answer. Having such a procedure would be an advantage.

Wolfram, however, has pointed out that the question of whether or not an effective procedure can be found corresponds to the question of whether there exists an effective algorithm for determining whether a universal Turing machine, given a random program and a random input, will cease its computations in finite time.[71] The two problems are equivalent, at least for the types of automata that can embed the code for universal Turing machines (we saw one such example in chapter 3 with the two-dimensional Life game). Wolfram believes that his one-dimensional class 4 automata, which exhibit a dynamic, "living" computational behavior, are clear candidates for the same degree of complexity. In this case, like Turing machines, they are also what mathematical logicians call "formally undecidable," and there is no guarantee that for a given automaton rule and a given start state there will be any adequate formula for predicting the developmental sequence of the automaton. They are *computationally irreducible*. The only way to predict the development would be to simulate the automaton itself; that is, to realize its own logical sequence step by step in a computer or using paper and pencil. This is certainly a somewhat pathological form of prediction: in fact, it is nearly the opposite. We risk having to wait 6,294,578 steps before we can be sure of what happens on the 6,294,579th step!

Translated into the language of biology, this means that we cannot predict very complex systems: to simulate them directly is the most effective means of determining their behavior. And this demands a model of the system with the same degree of complexity as the system itself, which undeniably makes it unmanageable as soon as we approach more realistic systems than those having a few rows of cells on a computer screen. For these kinds of complex systems there is no other way to predict their evolution than by tracking it. We must wait and see what happens. If this description also covers embryonic development, it means that the precise form of the final individual (the total phenotype) cannot be predicted, even with complete knowledge (which is itself purely hypothetical) of the total genotype and the epigenetic substrate.

This is a different, logically conditioned type of unpredictability than that which emerged out of Waddington's model of the uncertain bifurcating path through the epigenetic landscape. Waddington's model was vague (since then, many form models of greater precision and realism have been developed).[72] The model was continuous as to the embryo's possible state, and it was stochastic, in that real randomness can have decisive influence. In contrast, a cellular automata–based model of form creation is mathematically precise (one can then discuss its biological interpretation), discrete with reference to cell states, and deterministic, in that the current state of the automaton is entirely determined by the previous state, according to the rules. It would perhaps be closer to the truth to say that real living systems contain both aspects; they are both continuous and discrete (cells stick together in elastic wholes but are each distinct from one another). They are partly random and partly regular. The models thus complement each other.

BIRDS OF PASSAGE: ARTIFICIAL FLOCKING

Animals, too, can be modeled in different ways. Research into artificial life tends towards parallel computations over many smaller units, whether these are cellular automata, branch-like algorithms, or something completely different, like birds. Observing the flight of birds, one cannot avoid being enchanted by the way an entire flock of starlings, for example, has the phenomenal ability to coordinate the flight of all its individuals through a landscape that often contains many obstacles. Unlike geese and other birds that fly in v-formation, with a leader in front and two rows behind, a flock of starlings—and other bird flocks without fixed arrangements—seem to continuously solve an enormous coordination problem. Such flock behavior must demand nearly miraculous abilities. Who coordinates such a flock? There is clearly no lead bird, but only for the simple reason that it would not be immediately visible or otherwise identified in a stable fashion by the other birds.

Numerous small flying "information processors," purely local behavioral rules, together with collective behavior at the general level: it all sounds like a case for artificial life.

Craig Reynolds has shown that it is possible to achieve such bird behavior in a computer. Reynolds designed a computer program for flock behavior without wasting time on speculations about coordination by some kind of superior control. Instead, he defined three simple rules that each computer bird, which he called "boid," had to follow:

1. Maintain a minimum distance to other objects (including other boids) in the environment.
2. Seek to adjust speed according to the other boids in the immediate vicinity.
3. Maintain the position in the vicinity of that location considered by the boid to be the center of the flock (the "center of gravity" of the other boids nearby).

A migratory urge is built into the model and specified in terms of a global direction (as in "heading Z for the winter"), to make sure that the flock does not stay in the same location but continues forward. Apart from this urge, these are the only rules that govern each single bird's reaction to purely local events. What kind of ornithologist would have believed that this model would be sufficient to generate beautiful flight patterns through the computer landscape? Reynolds apparently did. Prior to formulating his three rules, he had spent many hours with other ornithologists, and with his binoculars observed real birds in nature.

The flock in the computer splits itself up into several smaller subflocks whenever it encounters obstacles. It reorganizes itself into larger subflocks, and reforms itself again when these barriers are passed. If boids are let go into each corner of the world in space, they will gather themselves together and create their own flock (see fig. 4.7). The global behavior of the flock is an emergent phenomenon in the model, for none of the rules for individual movements depend on global information. This is precisely the same in cellular automata. An interesting type of behavior is here simulated in a quite realistic way. Even if it is not life, it is certainly lifelike.

Chris Langton has no reservations. It is not just a simulation, that is, an imitation of a phenomenon. The individual boids in the model are certainly not real ones, they are but simulated birds, having no coherent physical structure, and they exist only within the informational structure of the computer. As far as their collective behavior is concerned, however, what boids do in the flock and what real, wild birds do when they gather, "ever hovering, hovering," are two examples of the same phenomenon, flocking. (And it is the level of the collective effects/emergent phenomena that constitutes decisive confirmation for Langton, and for the strong version of artificial life.) According to Langton, the flock behavior in Reynolds's model is bona fide lifelike behavior, genuine artificial life.[73]

FIGURE 4.7. A flock of "boids" avoids several pillars (from Craig Reynolds). The individual birds are simply simulations, but Langton considers that the total flock behavior can be considered genuine life, having emerged due to the boids' mutual interactions.

Artificial Evolution

The path of evolution is a long one, but life itself is short, an especially depressing fact for the biologist who dreams of carrying out controlled experiments with evolution. It is difficult to conclude such experiments in the time span of a single research career. Herein lies the biologist's affection for bacteria and fruit flies, or, if they are to resemble mammals: mice. All these organisms breed like rabbits. Experiments with evolution can thus be made within a microevolutionary time frame. Microevolution consists of the minor changes that occur over relatively short periods in a population's distribution of genetic types (from a few generations to several hundreds). Geneticists undertake selection experiments with fruit flies in cages. These function as biological models to represent the influence of natural selection on a population's composition during, for instance, fifty generations, i.e., within a microevolutionary perspective. The same is true for animal and plant breeders' deliberate selection, which creates artificially modified organisms, food staples like rye and wheat, beasts like Rottweilers, green nightmares like the Chrysanthemum and all the categories in between. These experiments with nature's gene pools have existed since the invention of agriculture, and the time scale is therefore longer. We can first speak of macroevolution when we deal with the creation of wholly new species or with the trends that can be deciphered in the geological layers of fossils. (A trend can be the reduction in the number of joints and their coalescence during the evolution of arthropods and insects, or the tendency of many animal nervous systems to bunch themselves together into brains.)

To date, no fruit-fly geneticist has been able to produce a new species of fruit fly in the laboratory.[74] Creation of a new animal species takes a long time. It is a quite poorly understood process in biology, and the facts are scanty; the paleontological data from fossils is often contradictory. Regrettably, many biologists have resigned themselves to a

quiet agnosticism: the obscurity of the matter and the short-
ness of life seem to indicate that we will never know much
about macroevolution.[75]

Nothing could be more wrong, however. First, the paleontological situation does not appear as depressing as first
thought; one can in fact observe a lot of specific patterns in
the fossils' geological record. Sometimes they indicate a
regular, gradual evolution; other times we can observe a
more punctuated sequence with many new species followed by long evolutionary pauses or "stasis" (there is,
however, great discussion regarding the interpretation of
these patterns; what is seen depends on the eye of the be-
holder).[76] More importantly, however, computer models of
macroevolution now make it possible to gain a better understanding of long-term evolutionary processes. There are,
to be sure, many obstacles to be overcome. The step from
the paleontologist's digging tool to the alifer's keyboard
and screen seems to be a large one. Some paleobiologists
(those who study past life-forms) may even feel threatened
by the alifers who never get their hands dirty. Some computer mathematicians cannot distinguish between basic biological concepts such as species or population. Cooperation is necessary, but it has hardly begun. Nevertheless,
several exciting models of evolutionary sequences have
been developed in which the computer crunches its way
through masses of calculations of long sequences of generations, selected partly because the environment, which consists of nature and of other species, itself evolves and
changes. A very dynamic situation indeed.

The Swedish physicist Kristian Lindgren at Chalmars
University of Technology in Gothenburg, and his colleague
Mats Nordahl (now at the Santa Fe Institute in New Mexico) have investigated the dynamics of a population of individuals who each have distinct strategies of survival. The
model they have constructed shows surprising patterns
that correspond to the descriptions of macroevolution as a
long march without special changes in the morphology of

the species (stasis), punctuated by short periods of violent evolution and creation of new species.[77]

Lindgren's model is quite simple. The individuals in the population have to survive in a game where the environment consists of other individuals. Each type of individual has a strategy encoded in its genome, and those who come off best are permitted to produce more offspring than the others. The game itself is a variation of the iterated Prisoner's Dilemma, where an individual—a prisoner in isolation—can cooperate with or defect on another individual (the other prisoner), and where the advantage (the individual's "fitness") of cooperating or defecting depends on what the other individual does.[78] The goal is to obtain a high number of points, corresponding to a reduced prison sentence. Both individuals obtain three points if they cooperate; one point each if both work against each other ("defect"); and if one chooses cooperation and the other defection, the latter obtains five points while the "exploited" and cooperative gets zero.

The game is not especially realistic biologically, but it is well known from game theory. It is easy to use it to define survival in an environment, with points accorded for the individual's strategy. And it is well suited as a simple framework for the core of the model, which concerns understanding the dynamics of natural selection. One could just as well have applied other criteria for determining whether a given strategy for survival was good or bad, and Lindgren has also attempted to allow the individuals to play other games against each other. But the important thing is that fitness depends upon what other strategies are on the market—that is, in the population. Ecologically speaking, an organism's fitness is profoundly dependent not only upon the physical aspects of the niche, but also on the other organisms dwelling in the habitat.

It is the *iterated* Prisoner's Dilemma game because the players are allowed to play against each other several times, and they can let their acts depend on what their opponent's acts were last time. Hence, the players can remember, and variations occur by introducing noise; that is, the players can make mistakes and execute a different act than the one they intended. Lindgren devised the system so as to let the genomes mutate (change at random places), so that new types of individuals and strategies appear.

In each playing round, the individuals decide if they will cooperate (C) or defect (D). The decisions depend on their genetically encoded strategy. A strategy is a rule that maps a given history (one's own and the opponent's earlier actions) onto a new action (C or D). For example, "If history is that he did D, then I did D, then he did C, then I will now do C."

The memory capacity of the organism becomes important here; if the size of the memory is two bits, for instance, one can only remember the opponent's last action and one's own before that. A sample strategy—a genotype—could read CDDC, which translates to the following phenotype rule for action:

> Cooperate [C] if the remembered history is DD (both defected the last time); defect [D] if the history is DC (I defected, then the opponent cooperated); defect [D] if the history is CD (I cooperated and then the opponent defected); cooperate [C] if history is CC (we both cooperated the last time).

The CDDC strategy is thus cooperative in the sense that it will generate cooperative behavior if both players did the same act the last time. (It is also stable against errors, but it can be exploited by a strategy DDDC.) For a two-bit memory, there are thus four possible histories and sixteen possible strategies; the possibilities grow exponentially if we allow more memory.

A point mutation changes one little part of the strategy, for instance CDDC → DDDC, and a gene-duplication mutation adds a whole copy of the genome to itself, for example CDDC → CDDCCDDC, which increases the memory capacity significantly, but gives the same strategy as phenotype. However, if genome duplication is followed by a point mutation, a quite new strategy will appear.

Lindgren's scenario contains one thousand individuals with a number of different strategies. A single generation is computed as follows: an individual with a given strategy plays against the 999 other individuals a large number of times. The same happens to the other individuals. Each player scores according to the rules above. Afterwards, the mean score of a strategy type (the group of individuals with the same genotype) is calculated and compared to the mean score of the total population: this comparison is expressed as the actual relative fitness of the genotype. Now the effect of natural selection is computed in terms of the increase or decrease in the number of individuals of a particular genotype in the following generation. The share of the particular genotype is thus determined by its actual share and how well it has scored in the foregoing round (i.e., its fitness). Finally, this next generation is formed by allowing a small chance for mutation: the mutation frequency is two changes for every ten thousand perfect copies of the genome.

The game is played for hundreds of generations, and the number of computations is enormous, thus requiring the use of a computer. At the start, the population is dominated by a few genotypes with simple strategies that are released to fight one another: the bad DD, who always defect, exploit both CC and CD, and therefore increase in number. But then the DC strategy begins to grow. This strategy repeats the opponent's previous behavior; that is, it cooperates if the opponent cooperated last time, otherwise it defects. It is called the "tit-for-tat" strategy, and after a few hundred generations it has decimated so many of the annoying DD that the environment now becomes friendlier. Then the CC, who always cooperate, begin to increase their share in the population, for CC is not exploited by a tit-for-tat strategy (see fig. 4.8).

If the game is simulated over longer periods, some extremely interesting patterns evolve on the curves generated

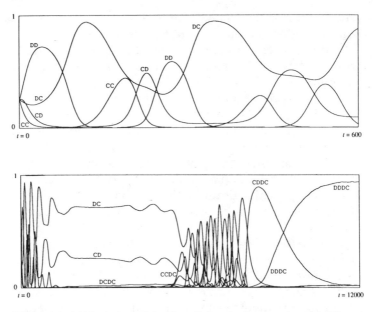

FIGURE 4.8. (a) Evolution of the first six hundred generations of different genotypes, which vary in number (indicated by the height of each curve). Notice that DD (always defect) dominates at the beginning, but is soon ousted by DC (tit for tat) such that CC (always cooperate) and CD (anti–tit for tat) can again increase. This subsequently produces a situation whereby DD obtains fertile conditions for growth; (b) Simulation of the game over a longer time period, where mutants emerge with larger memory. Long periods of stability in the composition of genotypes is replaced by revolutions: rapid evolution of new types.

by the computer (fig. 4.8b). After the first fluctuations recede, there occurs a long period with stasis, an interaction between tit-for-tat and anti-tit-for-tat (DC and CD), which stabilizes their number. This harmony is suddenly punctured when the mutants, with their greater memory, come onto the scene. The situation becomes extremely unstable as several strategies compete. A mass extinction of types occurs, and the winner cannot be predicted. After a period of rapid evolution and large variations, the system again stabilizes itself when the strategy CDDC increases in number.

This strategy is quite friendly but can be exploited by one of its mutants, DDDC. As soon as it emerges, the days of friendliness are numbered. The DDDC grows slowly, and after about ten thousand generations the population enters a long, monotonous state of very slow evolution dominated by this new, rather egocentric strategy.

Now it might be thought that DDDC would constitute a sort of evolutionarily stable strategy, which does not allow itself to be invaded by others. But nothing, including stability, lasts long in evolution, and after several thousand generations new mutants appear, which generate new, powerful disturbances in the system. After about twenty-four thousand generations two strategies (with still greater memory) emerge that together come to dominate the scene: CDDCDDDC and DDDCCDDC. We might call them Adam and Eve. These two strategies cannot cope with mistakes if they play against one of their own, but if mistakes occur when an Adam plays against an Eve, after a short period of defection they can revert to mutual cooperation. Lindgren calls this "mutualism": it is ecology's label for two species that live in symbiosis for the advantage of both. It is a beautiful example of one of the many emergent phenomena that can be observed in models of this type: coevolution and cooperative behavior.

Hence, we can decipher several exciting patterns from the evolution of complex systems: stasis, punctuated equilibrium, varying speeds of evolution, mass extinctions, coevolution, and growing complexity of the behavior patterns and strategies that survive. This is especially interesting in view of the simplicity of the game's starting conditions. The complex behavior is driven forward by the dynamic system's own inner development. For paleobiologists, there is an important moral to this story: they no longer need to search for or invent "external catastrophes" like volcanos or meteors crashing to earth in order to explain violent changes in an ecological system.[79] Even though the biological interpretations of Lindgren's model can be debated,[80] it demonstrates that stochastic dynamic systems (with a built-in fitness function and natural selection) can exhibit immensely varied behavior based purely on their inherent factors. Moreover, natural ecosystems must be considered to have a degree of complexity and a corresponding plethora of nonlinear relations at least as extensive as the computerized system models.

In other words, catastrophes do not always come from outside, but can appear as small cracks in the dynamics within which normal systems operate.

Venus's Artificial Chemistry

In simulating evolution, we may speak of the evolution of something, in this case types of individuals. But can we also simulate the evolution of evolution itself? How did it all start? How did the evolutionary game itself evolve? Answering this question is really another way of trying to find a solution to the problem of the origin of life, for life as such must also be understood as a result of evolutionary processes. Ever since the Russian biochemist A. I. Oparin's theory in 1924, and the experiments of Stanley Miller and L. E. Orgel in the 1950s and onward, biologists have sought to understand the emergence of life as a gradual sequence of complicated chemical processes within a wet chemistry consisting of some primitive versions of the component parts of the living cell. Four billion years ago, the seas were a powerful soup of organic molecules, whose basic structures consisted of chains of carbon atoms. If a similar soup can be made in the laboratory and if we can trace the same selective chemical transformations that led to the chaining together of simpler molecules into longer and longer macromolecules—that is, proteins, nucleic acids, flexible membranous structures, and much more—then we might eventually find the missing link between soup and cell; in other words, we could fill in all that is missing in the transition from a more or less composed, vitamin-enriched synthetic laboratory soup to the incredibly structurally rich and functional wonders that characterize even the simplest living cell.

Wet chemistry and biochemistry are difficult fields. Moreover, the enormous complexity of all possible chemical reactions on earth often requires that one try to find a single, predefined path to life. However, once the methods of artificial life are discovered, the possibilities for self-organization in artificial chemistry must also be investigated. This self-organization may be based on some simple atomic instructions that can enter into "chemical" reactions with each other. A group of young researchers at the Center for Modeling, Nonlinear Dynamics and Irreversible Thermodynamics at Denmark's Technical University and at the Theoretical Division at Los Alamos National Laboratory in New Mexico has proposed this method as well suited to investigating the open, indeterminate character of the process that brought about life. The method does not freeze into one specific path among the many possible paths from chaos to organized, cooperative life-forms. The hope is that the procedural structures that pop up in the artificial space of possibility can be connected to certain similar processes in the space of genuine chemical reactions. Steen Rasmussen, now working at the Los Alamos National Laboratory, believes that if we want to learn how an evolutionary system chooses its own paths, the methods of artificial-life research are the most promising when compared to controlled molecular biology experiments that never become "experiments where the system is simply released and organizes itself into something 'alive.' "[81]

What kind of artificial chemistry have Rasmussen and his Danish colleagues created? It consists of modifications of the elementary instructions of the machine code that is the basis of every normal computer. The idea is to allow the elementary instructions to interact among themselves and to exploit the computational resources in the computer's memory, and then to examine whether many various small instructions are able to organize themselves into interesting patterns. The strategy is reminiscent of Ray's model (see n.

31), but where Ray designed an actual organism from the start and allowed it to evolve, Rasmussen and his colleagues begin with a more raw, disordered chemical chaos. Let us examine the "substance" and the "molecules" in the model developed by Rasmussen and his colleagues. The computational "fluid" consists of a *core*: a long sequence of 3,584 addresses, each of which either contains an instruction or is empty. There are ten different instructions, and we can simply note that like other machine-code instructions, these concern managing data in the different addresses of the machine's active memory. MOV(A,B), for example, is an instruction that moves the information at address A over into B. The instruction ADD(A,B) adds the content of A to B and puts the result into B. The activity-creating instruction, SPL(B) divides the execution between B and the next instruction, providing extra computational resources in a prespecified instruction. The instruction JMP(A) moves the pointer to the address A. (The pointer is the function that determines whether an instruction should be executed or not: an instruction will be executed only if an address has both a pointer and sufficient computational space in its neighboring spaces.)

The physical computer is not allowed to become the core: that would be completely unmanageable. Rather the model is embedded into a program that simulates a virtual machine's core in which evolution takes place (see chapter 2 on computational containment). The program is named VENUS, which refers to both the goddess born from sea foam and to "Virtual Evolutionary Non-deterministic Universe Simulator." The nondeterministic label derives from a built-in small chance for mutations in connection with the execution of the MOV instruction. The VENUS program contains several other smaller programs that can react among each other, purely locally, during the consumption of a limited amount of energy and fertilizer in the form of the available computation time and space. (The metaphors are not mine; they lay behind the research group's development of the entire program and form part of the presentation of the results from the experiments they have executed with VENUS.)

After many attempts and after examining the evolution of the cores—here "evolution" appears as kilometer after kilometer of printout—the group began to observe signs of the emergence of distinct structures. Under favorable conditions ("jungle") where the core during the entire period contains an infinite array of computational and reaction possibilities, cooperative structures that had not been predicted from the start popped up. An example of such a simple, though very stable, cooperation is the "SPL-MOV organism." The SPL instruction attempts to divide calculation time between itself and the MOV instruction, while the MOV instruction seeks to copy itself and SPL. A SUB instruction can be integrated into the organism, which makes the allocation of the computational resources even more effective. Interpreted in the language of wet chemistry, Rasmussen here depicts a self-reproducing network that evolves by incorporating new chemical reaction paths. This is made possible via the diffusion of new materials into the area.

Has artificial chemistry then created the basis for authentic artificial life? Rasmussen does not believe that anyone has succeeded in creating processes that we could term living.[82] Nonetheless, VENUS has evolved interesting, stable patterns within the total space of possibilities. It can teach us something about evolution and self-organization: the historical events, which are the details of the noise that affects the system via the random mutations, have considerable significance for the system's choice of evolutionary paths. Some of these are boring and monotonous, while others evolve into exciting "life-forms," such as the partly stable cooperative structures that can come to dominate most of the core.

Optimal environmental conditions are not enough to insure the development of cooperative structures in the model, because randomness and historical events play a central role as well. During the field studies of VENUS, the group identified evolutionary "epochs,"—i.e., various macroscopic conditions where the total system exhibits different properties—from monotony, through phases with simple instruction loops, to phases where the core is populated by special types of instructions, to the evolution of cooperating programs. However, we cannot speak of genuine open-ended evolution, for the system always seems to be caught in one attractor or another, where the condition is thereafter a very specific one.

Yet the significance of the initial conditions is great, and it might well point toward the system being chaotic (inasmuch as great sensitivity to initial conditions is the most important sign of a chaotic system). It is hardly the second creation of life. And yet, "We have set up a little universe, and we did not know in advance what would happen," as one of the team members, Carsten Knudsen, remarked. "It is a kind of life."[83] Rasmussen formulates it in this way: "The evolution which brought forth our SPL-MOV-organism is just as correct or 'real' as biological evolution. However, it is still very far from the primitive properties which this computer structure possesses to the intricate interaction of functions which constitutes a living biological cell. But I would assert that it is a question of quantity and not quality."[84]

GENETIC NETWORKS

Stuart A. Kauffman from the University of Pennsylvania is one of the top researchers in theoretical biology. He belongs to the generation that connects the classical theoretical biology of the middle of the century—people like C. H. Waddington, J. Needham, J. H. Woodger, E. Mayr, and others—to contemporary trends within artificial life, chaotic systems, and complex dynamical systems in physics and mathematics.[85] With a background in medicine and biochemistry, Kauffman has long been interested in how a cell achieves its individuality; in other words, how the 210 different types of cells of which our body is constructed differentiate themselves during embryonic development and growth. Nearly all cells (the red corpuscles are an exception) contain the same genes. The difference depends on what kind of genes are active in the individual cell types. The gene for the creation of eye color is also found in the liver cell, but is somehow turned off there.

We know that the process is genetically controlled, but how? There are other open questions as well. What is a cell type basically? How is it differentiated in a stable fashion? For instance, can a cell that normally creates connective tissue become under certain circumstances a cell type that could create cartilage or bone tissue? What is the developmental path of a cell type? When cells are differentiated, they will typically first differentiate themselves into a few cell types, which in turn differentiate into a few other types, and so on. Why does there not occur more direct differentiation into the various final range of the 210 types? What overlaps exist in the genetic expression in the various cell types? In other words, what turned-on genes are found in both cells? If a gene is eliminated, how many genes then change their patterns of expression? Even though we know much about the regulation of individual genes, we know very little about the interplay among the many genes in the cell.

Kauffman has never believed that these kind of questions could be explained solely by referring to the structure at the single-gene level or by simply invoking natural selection. Evolution, he believes, is a combination of natural selection and other, internal factors derived from the ability of complex systems to spontaneously self-organize. Every human cell contains about one hundred thousand genes, including an unknown number of regulating genes that can turn each other on or off in an enormously complicated network of connections. It seems to be the perfect recipe for confusion. One might expect that the system would constantly go haywire, but it does not, aside perhaps from cancer cells. The genome organizes itself into stable patterns of activity, corresponding to distinct cell types: a muscle cell, a skin cell, a mucous membrane cell, a brain cell, etc.

In contrast to the current, hard, molecular biology method—to concentrate upon one specific cell type, characterize one of its specific proteins, isolate and purify it, map the gene that codes for the protein, sequence the gene, investigate which other genes, DNA-binding proteins, or other factors regulate its expression—Kauffman has chosen an approach that in a certain sense formalizes the total problem that had puzzled him from the start. He has built a mathematical model of a genetically regulatory system as a logical network of connections: gene A turns on gene B, which in turn inhibits gene C, which then works to activate gene D, etc.[86] There can be hundreds of genes in the model. We know that one gene (or one regulatory locus on the DNA) has a limited amount of regulatory input; for example, the lac-operator is controlled by at least two variables: the repressor molecule that binds to the operator site on the DNA, and the allolactose, which binds to the repressor and pushes it off the operator. Since the regulatory molecules are themselves often coded by another gene, one can view the genes as directly affecting each other. In its simplest form, Kauffman's model has the character of a binary (on/off) automaton or a so-called Boolean logic network.

Logic lies in the modeling process. Imagine an extremely simple network with only three genes (fig. 4.9a) each of which can be turned on or off depending on the previous condition of the other two genes. Gene 1 is controlled by the Boolean AND function; that is, its activity demands the activity of both genes 2 and 3. Gene 2 is controlled by the Boolean OR function; that is, its activity demands the activity of gene 1 or gene 2 or both of them. Gene 3 is also governed by the Boolean OR. At any point in time, the actual activity pattern of on or off genes defines a state of the total genetic network. In this example eight different states exist, and these constitute the *state space* for the total genetic network. For each time step the system passes from one state to the next; hence, each gene is changed according to the activities of its input when it passes into the next time interval (see fig. 4.9b). Over a certain period, the system passes through a sequence, or a *trajectory* of states (fig. 4.9c). Since there are a finite number of states, the system will pass through states it has passed through previously. The system is deterministic and will therefore flow around in the same closed sequence of states, called a *state cycle*.

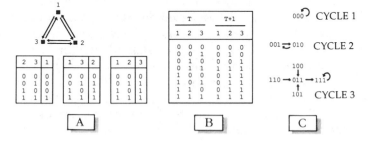

FIGURE 4.9. (a) Three genes—1, 2, and 3—are each regulated by the two others. Gene 1 is turned on in the next time interval only if genes 2 and 3 are active (see the bottom line in the first small table above) in the present time interval, corresponding to the logical AND function. Gene 2 is activated in the next time interval if either gene 1 or gene 3 or both of them are active (corresponding to OR). In the same fashion, gene 3 is activated if either gene 1 or gene 2 or both are active (OR). (b) The Boolean rules in (a) are rewritten. In the left half are shown the eight combinations of activity states of the three genes at time T, in the right half the next state (time T+1) of the genes. (c) The sequence of states undergone by the system in time, as it can be inferred from figure B. ("101" is read as "gene 1, 2, and 3 are, respectively, on, off, and on.") There are three state cycles. Cycles 1 and 3 are steady states of constant gene activity. Cycle 1 is a two-state oscillation (after Kauffman 1987: *BioEssays* 6:82–87).

Figure 4.9c shows the dynamic behavior of this little network. Each of the states (e.g., 001) belongs to a trajectory that leads to (or is on) a cycle (in this case the cycle 001 → 010 → 001 → . . . , where the first and third cycles constitute a single point in state space). When the network is larger, many different subsets of states are created where each set

flows in the direction of its own cycle, which therefore operates as an attractor. A set of states flowing toward one state
cycle constitutes a *basin of attraction* in the state space. One
can view the course of events as a sequence of states precisely like the ball in Waddington's epigenetic landscape:
the ball rolls down toward its attractor. The network will
tend to remain in these attractive parts of the total state
space, not because the network is equipped with any purposeful intentions, but because it can do nothing else; its
state will be captured by an attractor. The little network has
three attractors (fig. 4.9c), one swinging back and forth between two states. A larger network will have many alternative attractors, but they will still each divide the total state
space into their respective corners or basins, which pull an
activity pattern into a certain trajectory and maintain it to
an attractor.

What, then, is the bio-logic, the biological content of the
Boolean network? Kauffman interprets each state-cycle attractor as a cell type. Our mininetwork can thus realize
three different types of cells with the same genes; each type
is determined by a characteristic distribution of on and off
genes. It is a good interpretation (and a good model), for
cell types are indeed stable patterns of gene activity, not
random patterns. Besides, healthy cell types do not have all
possible intermediary forms but divide the continuum of
abstract possibilities into discrete natural classes. (We might
imagine that all conceivable intermediate forms between a
liver cell and a nerve cell could be realized, but in fact they
do not exist). Kauffman has thereby achieved a useful language for talking about cell types and modeling their development.

Now Kauffman can define a more quantitative, realistic
model and ask, "What can we say about the dynamic behavior of a large genetic network of one hundred thousand
genes, where each gene is regulated by an input of two
other genes, and what becomes of the regulatory architecture of such a network?"

This is an insane question, for there are nearly an infinite number of ways in which one hundred thousand different genes can be internally connected. Therefore, Kauffman allows a computer to roll the dice about the detailed connections, so that the networks are constructed randomly; the only factors specified are the total number of genes and the number of inputs per gene.

Surprisingly enough, even such a randomly hooked-up network can exhibit behavior in a biologically well-ordered fashion. It appears as if the creation of a relatively small number of attractors, corresponding to cell types, is an unavoidable property of such a dynamic system, despite the fact that it starts entirely randomly. Kauffman's simulations show that the number of attractors is given roughly by the square root of the number of genes, if each gene is regulated by two others. That is, one hundred thousand genes yield about 320 attractors, ten thousand genes about one hundred attractors. It is not such a poor prediction when one compares it with real animals. Humans, it will be recalled, have about 210 cell types and about one hundred thousand genes; simpler organisms have correspondingly fewer genes and cell types—a similar quadratic relationship exists between the amount of DNA per cell and the number of cell types for species of bacteria, algae, fungus, worms, and medusas. If Kauffman allows each gene to be regulated by not just two but every other gene, so that the number of genes and the number of regulatory inputs per gene is the same (a very unrealistic model), the system will become extremely unstable: the smallest disturbances can push the network toward another attractor.

Models of this type can help answer the kinds of questions Kauffman has posed about evolution. A realistic model points toward genetic homeostasis—i.e., stability in the face of external disturbances—being built into the system. It does not necessarily demand the external tinkering of natural selection. It is also striking that a given cell type in the model can only differentiate into a few other cell types (nearby attractors). The multicellular organism's typical branching patterns of cell differentiation seem to be built in as a necessary result of the system's structure. The branching pattern, like a family tree, shows the origin of the various cell types in embryo development. A nerve cell in the eye, for example, originates from a neural stem cell, which again comes from an epidermal cell, the cell lining of which can be traced all the way back to the fertilized egg cell. It is thus also possible to understand that a fertilized egg cell cannot directly reach all possible cell types in the state space of cellular differentiation: the attractors form a landscape of barriers consisting of narrow mountain passes, blind alleys, and high peaks. During the course of evolution, mutations may induce the choice of new pathways of cell differentiation; the old pathways remain as silent, potential alternatives. This creates the possibility for the emergence of atavistic cell types (such as the teeth that occasionally appear in a hen's beak) or hitherto unrealized cell types that exhibit cancer-like behavior.

The above-mentioned properties in Kauffman's model do not appear in every network. They are mathematically nontrivial. They require, among other things, a small amount of regulatory input per cell (which is *canalized*, i.e., one of the gene's states is alone determined by one of the regulatory inputs). If the number of inputs is greater than four or five, several of the properties disappear. The conclusion is that complex systems of the types Kauffman has modeled exhibit a series of internal, generative properties created out of the dynamics of the evolving system itself.

It was the geneticist Dobzhansky who once said that nothing in biology makes sense except in light of evolution. He was certainly correct. Today we know more about evolution and self-organization in physical, chemical, biological, and computational systems, and we can therefore apply a broader perspective on evolution. Neo-Darwinists have occasionally misinterpreted Dobzhansky as claiming every evolutionary phenomenon to be explainable in terms of natural selection. Thanks to the work of Kauffman and others, we can see that a whole range of evolutionary phenomena are internal, inherent properties of self-organized systems, and that natural selection need not lie behind all the structures that biologists observe in nature. Natural selection—along with other factors—is probably an essential cocreative factor.[87] Moreover, selection can now also be understood as the means by which a dynamic system chooses its own trajectories or evolutionary paths.

Chapter Five

THE ECOLOGY OF COMPUTATION

THE STRANGE world of a-life is not simply artificial in the sense of being artificially created, a cultural product of human ingenuity. It is also a world containing several not yet realized possibilities, a world of potentials that we can hardly imagine, and in this sense it is a virtual world.[88] The world of a-life is at the same time an immaterial world, realized as a flow of informational structures within the machine. It is a world that seems to lack any kind of material existence.

Up to now we have restricted ourselves to simulations of specific models and to constructions—realizations—of abstract computational systems such as cellular automata within the computer. Could the programs of alifers be inserted into actual machines and equipped with sensors and movement apparatus? Could they be realized as material, physical beings? What is the relationship between robots, artificial life, and the systems already developed by researchers in artificial intelligence?

ROBOTS AND ANIMATS

The robots that now work on car-assembly lines and in several other automated production processes are the result of decades of research within cybernetics, artificial intelligence, and several other engineering-related disciplines. It is immensely difficult to build a robot that can carry out anything more complicated than a few simple operations (such as separating objects using simple, predetermined patterns and putting them into their respective boxes). To design and construct a robot that can move itself independently in an environment—to move through a room without bumping into the furniture or other moving objects such as humans—causes trouble for AI research groups. A general goal such as the development of watchful, self-moving machines is pursued by breaking complex capacities down into functional subtasks, each of which is in turn broken down into simpler tasks and movements. These tasks could involve the capacity to recognize complex visual stimuli, to interpret the environment as a scene for possible actions, to analyze these actions and integrate them into the system's high-level hierarchy of goals, to generate a plan of action that could dynamically account for change in objects' positions, to convert the plan to motor commands, etc.

It is often the case that such a task fragments itself into separate subproblems that become independent research areas within artificial intelligence. Machine vision is a successful example of one such area. Artificial intelligence has many such areas where specific parts of an intelligent system's abilities are modeled; e.g., search, representation, play simulation, expert systems, proof of theorems, pattern recognition, inference, etc. Specific systems can be developed for solving specific tasks, but it is difficult to integrate these intelligent tools into a coherent whole.[89]

AI researcher Richard K. Belew, who has long worked with machine learning, belongs to the moderate group of AI specialists who understand the limitations of the purely logical approach to constructing thinking machines. Below is fascinated by the new biologically inspired design principles that will apparently form the basis of the architecture of hardware and especially software in the computers of the future. Neural networks are one example of the biological-programming principles. In addition, cellular automata, L-systems and so-called genetic algorithms (see below) can be viewed in terms of biological computation.

Below hopes that the alifers will be able to find new ways of viewing the design problems connected with robotics. Organisms were certainly not released into nature only after they were constructed from functionally perfect designs. Evolution operates like a tinker who fixes a broken machine using the materials at hand. Not every design is a good design; many are called, but few survive. The robot builders might learn something by studying the evolutionary game. Instead of constructing expensive, complicated machines designed for a limited number of well-defined tasks, we might instead build a whole flock of cheap, simple, and perhaps rather unpredictable machines and allow them to evolve gradually. Rodney Brooks, working in the Robot Laboratory at the Massachusetts Institute of Technology, has proposed to his astounded colleagues the slogan "quick, cheap, and out of control"; it is a way of thinking that could bring the laboratory's methods in line with the evolution of living systems.

Many animals can execute operations reminiscent of logical inferences where they extrapolate from a single instance to a general rule or habit (induction). Perhaps higher animals have some kind of map or model of the environment built into a small gray mass of nerve networks. If robots must be autonomous, companionable, and possess at least a simple, primitive form of a sense of occasion, a "Fingerspitzengefühl," we must build into them a theoretical understanding of adaptive and learning behavior. Perhaps they will become kinds of robots other than those we know from today's factory floors or science-fiction films. They will probably come to resemble animals: they will become *animats*, without necessarily behaving in an animal-like or brutish fashion.

A-LIFE AS PROTOFORM FOR A-INTELLIGENCE

This new movement within a-life may lead to a considerable renewal for the field called cognitive science, which studies the thought processes of human and artificial systems. Perhaps we may not be able to crack the cognition problem or create an understandable scientific theory of thought until we understand what it means to say that something is alive. Life comes before intelligence, and a-life can show itself to come before real AI. The entire question of adaptation and development comprises the core of a-life, but it is equally important for intelligent systems. In both cases we are speaking of systems that determine their own behavior based on the information structures they contain. The problem with AI research is that we have sprung directly into the most complex example of intelligence; namely, human intelligence. We have been cheated by the fact that computers can do some things that people find difficult.

If we take a dead bird and throw it up in the air, its path describes a parabola, in conformity with the laws of motion (ignoring air resistance and the earth's curvature). Take a living bird and throw it up it the air, and something entirely different happens. Fundamentally, it is a matter of understanding how, given a physical universe dominated by matter and energy, systems can emerge that determine their own behavior by means of information or computation.

Artificial life and artificial intelligence are already on the way to constituting a continuum of projects that attempt to model adaptive, learning, and cognitive abilities in all the varying degrees of complexity we know from biology and psychology. A-life can be viewed as a science that concerns itself with the minimal level for thinking and the lower limit for sign manipulation and computation ("low" is not meant in the sense of a crude, egotistical personality calculating to one's own advantage, but in a computational-logical sense as the processing of signs and symbols): how simple can a physical system be before we can call it computational (and alive)?

This a-life–AI thesis can be formulated as follows: "The dumbest smart thing you can do is to stay alive."[90] Animals do it. We do it, too. Intelligence is not a question of either/or, but rather of different ways of staying alive. Organisms' apparently highly evolved, coherent behavior can often be explained as simple reciprocal interactions with a rich and varied environment. Much of the complexity seems to lie in the milieu. Think of an ant: it crawls around on the forest floor, carefully avoiding large barriers, but it must take minor detours in order to create space for dragging home a pine needle to its nest. The ant pauses from his work and exchanges information with a fellow ant. It usually has an especially complex route, but the ant as a behavioral system is quite simple—the complexity is to a great degree a reflection of the environment in which it finds itself. The point here is that if we have visions of constructing serviceable, sociable robots or such things, we must first discover the minimal procedures that enable an animal to cope minimally with its nearest surroundings. This does not sound like much, but it is. An ant can never imagine what it might meet on its path. Openness, adaptability and flexibility become more important than having a ready response to every conceivable situation, regardless of whether the response can be coded as a frame, a scheme, a script, or as one of the other AI techniques of representing knowledge. "Go to the ant, thou sluggard" (but remain ignorant)!

GENETIC ALGORITHMS

An animal species' particular way of coping with a demanding environment can be considered evolution's way of solving a problem. A group of researchers within artificial intelligence has utilized evolutionary principles in attempting to teach machines to solve problems. Evolutionary processes can be simulated and used as a kind of adaptive algorithm. An algorithm is a sequence of computational operations needed to solve a problem, and a program consists of the description of one or more algorithms. The special feature of evolutionary algorithms lies in the interaction between different variants of "individuals," where each individual corresponds to an algorithm that solves a given problem. The individual's relative success with respect to the solution is calculated and considered the "fitness" that is used to selectively reproduce the offspring algorithms that resemble, but are not wholly identical with, their ancestors.

John H. Holland's genetic algorithms, invented in the 1960s, have become quite well known.[91] They are especially effective because Holland was not content to view mutations as the only source of changes in the algorithms. Mutations very seldom improve the algorithms because of their random character (they correspond to a random search among parent generations' achievements), just as organisms are seldom improved by a mutation. Hence, in addition to mutation, Holland utilized genetic recombination: in biological cells the genetic material can be recombined in an infinite number of ways: during the generation of sperm and egg cells from germ cells, some of the germ cell's chromosomes may cross over so that for example, a chromosome in the egg cell consists of some segments that derive from the germ cell's maternal progenitor and some which derive from its paternal progenitor (recall that for each chromosome pair, one chromosome is inherited from the mother and the other from the father). Crossing over recombines the genes, or the instructions (fig. 5.1). The variants of the chromosome (in living cells) or the algorithm (in Holland's computer), which are the result of a crossover, have a greater chance of successful performance than do those variants that are the results of random mutations. Evolution is hitched to a wagon of problem solving according to the following procedure, where the point of departure is a population of simple algorithms that can provide only poor, approximate solutions to a problem:

1. Select program pairs on the basis of how well they have solved the task (one can thereby measure fitness). The better the solution, the greater the chance of being chosen.
2. Apply the genetic operator (crossing over, eventually combined with a small chance of mutation) to the selected program pairs in order to create offspring in the next generation.
3. Replace the least successful programs with the offspring created in step two, and repeat the process.

Empirical investigations indicate that this crossing-over scheme operates especially well on problems that program-

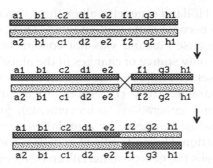

FIGURE 5.1. The principle of crossing over. Biologically: two homologous chromosomes cross over (i.e., their genetic loci correspond to each other, and they stem from, respectively, the father and mother progenitor). The genes (letters) in various editions, the so-called alleles (numbers), are recombined in new ways due to crossing over. Closely neighboring genes that eventually have operated well together are likely to remain together (coupling). Algorithmically, the instructions in the two programs are put together in a similar manner.

mers otherwise regard as genuinely difficult. The genetic algorithms are able to exploit the population's experience in an optimal manner, such that it is maintained in the distribution of alleles or program variants in the population. An example of a computationally very difficult task is the traveling salesman's problem, which consists of finding the shortest possible route between n towns, each of which must be visited only once. It sounds like a simple optimization problem, but if n is a large number, the necessary computations increase explosively. If we try for fun to find the shortest route between all the state capitals of the United States, we quickly realize that even this simple combinatorial problem requires a considerable amount of calculation. The genetic algorithm cannot solve the problem exactly, but its small-step tinkering method insures that this method, given the implementation of the algorithm, can proceed forward relatively easily and indicate some of the good routes, which are presumably close to the routes that in absolute terms are the shortest possible ones.

Research in artificial life has now entered a phase in which attempts are made to combine the various computational principles applied in nature in genetically and neurally based learning. Richard K. Belew, John McInerney, and Nicolaus Schraudolf of the University of California at San Diego have created hybrid combinations of neural networks and genetic algorithms, as they express it, "simply because nature has done so."[92] Yet even though nature uses both population-genetic and neurological methods, it is not always so simple to determine their internal relation, the interface. One problem lies in determining how explicitly the network's structure and the weights of the individual connections in the net must be encoded into the genetic algorithm that generates a new net in each generation. This is difficult to tell in advance, and demands some trial and error, just as with biological evolution. Evolution works in a nearsighted way, but in contrast to human designs it has plenty of time and the widest possible perspective. Evolution often finds applications for unforeseen side effects, which the human system-builders attempt to avoid. Without dwelling on technical details, it now seems that genetic algorithms' global form of exploitation and experience gathering constitutes a promising supplement to the neural networks' more local searches for structures in the data input they receive.

BIOCHIPS

We have gained an overview of some of the ways in which biology can be joined with computer science, mathematics, and physics. But what about the "biological computer"? Will today's silicon chips give way to a computer of tomorrow containing chips of biological material, *biochips*? Will the biochip bring the machine to life?

This is certainly the implication if we peruse headlines such as "Living Computer On Its Way: New Super-Chips Work with Enzymes and Acids," or "Biochip Can Replace Silicon." Headlines like these periodically appear in newspapers and computer magazines.[93] I do not believe that the biochip will attain any direct significance for artificial life, at least in the next decade. Biochip research seeks to make the computer's hardware more effective and press more memory and computing capacity into a smaller space. The silicon chip's possibilities for higher efficiency are far from exhausted. Nevertheless, it is calculated that the limit will be reached by the year 2000. The tiny holes in the silicon, where the memory is stored, will be reduced to 0.1 micrometers, equivalent to about four silicon atoms. This is not very large, but the margin of error must be maintained at a low level. Instead of silicon (but without entirely replacing it), we are now investigating the possibilities of using organic molecules. The research is taking place in deepest secrecy in laboratories belonging to the U.S. Navy and in giant Japanese research centers, so we do not often hear about it. However, it should be noted that the organic molecules being used are polymers, long chains of carbon-based molecules. Polymers are no more organic than the molecules that make up a plastic bag. The biochip, then, is not alive. It is not anything that will make the computer soft, cuddly, long-haired, or biological.

For this reason, most researchers prefer to call the biochip a Molecular Electronic Device, or MED. Nevertheless, there are several other projects in Japan that more directly investigate the possibilities—at the level of hardware—to connect computers with biology. Can a neural network, for example, be grown in a test tube? One could certainly grow cell cultures of nerve cells, and the possibility of more controlled growth and coupling to microelectronic circulation is also being investigated. This could be the beginning of the development of a neurochip. Nevertheless, many engineers in the mainstream computer hardware industry are asking what it is that biological molecules have that cannot be imitated by semiconductors, superconductors, and other synthetic materials. As long as it concerns effective, controllable computation, this question is difficult to answer.

The Japanese Bio-electronic Device Project is currently conducting research intended to discover information processes in living organisms.[94] This research involves visual information processing and movement control in the brain, and the development of other forms of molecular electronic devices such as biosensors. A biosensor is an apparatus capable of recognizing distinct smells, measuring the concentration of certain materials in the blood, or processing optical information at the molecular level. Biosensors are an example of highly advanced technology on a small scale with many applications both in industry and the health sector.

Could it be possible to utilize DNA as the basic material of a real biological chip? DNA stores information in living cells, and with our present-day DNA technology, and our gene-splicing ability to use enzymes to cut DNA at specific places, it should not be difficult. DNA has been considered a candidate for the biochip, of course, but was quickly rejected: what is needed are simple, organic molecules that can change between two states, and where this change of states can be controlled using simple external means. DNA is an enormously heavy and relatively stable molecule, and to imprint or "write" something into it—to change the information coded by the sequence of four different bases, which themselves constitute an integral part of the molecule—demands a highly complex enzymatic machinery that cannot be found anywhere else besides the living cell. One can certainly artificially synthesize smaller, shorter RNA and DNA molecules, and this synthesis (of oligonucleotides) has even been automated. Molecular biologists can now purchase a "gene machine" that produces small fragments of genes. Also the reading—i.e., the sequencing or mapping of the genetic material—will soon be automated; it is a spin-off of the mapping of the entire human genome. However, this is a quite different process than utilizing DNA technically as a bit container similar to a silicon chip or magnetic tape. A DNA chip will probably never be produced.

A Computation, Please

The images of chaos and fractals comprise a world of fascinating patterns embedded in other patterns. The soft, spiraling forms that pop up in the Mandelbrot set appear organic. Some of these patterns have also been given lyrical names that arouse our biological curiosity. One can gaze down into Sea Horse Valley and follow how the computer images sharply zoom in on forms like mini-Mandelbrots, branching filaments, and "elephants" marching down the heart-shaped cardioid curve. Aided by the computer, mathematics suddenly allows us to see a wealth of structures, some of them horrible, others beautiful and lifelike.

We know fully well that the sea horses in the Mandelbrot set and its infinity of other figures are generated by an extremely simple formula that a computer has crunched through for a myriad of values. We also know that real sea horses are a special breed of fish, born from other, real sea horses and constructed of cells with their own wet biochemistry. The similarity between real and computer sea horses is therefore superficial. Fractal sea horses are children of calculations, just like the computer organisms we have seen. Will our experiences with the real and the computed animals ever be able to confront each other?

The strong thesis about artificial life that contains all seven commandments (see pp. 17–20) can be considered a contribution to a new computational natural philosophy where the concept of computation in physical and biological systems occupies a central place. If natural processes are essentially computational ones, we can also study them by producing the same or similar processes in a computer. The processes can be abstracted from their material substratum. In its radical form, this natural philosophy thus becomes a program for a form of science that locates itself somewhere other than in physics, biology, and psychology, which are based upon experiences of concrete types of systems. We are now able to construct objects that transcend those that we normally experience in the world, but which can nevertheless be related to them.

The concept of computation is a double-edged sword: on the one hand we view it as something quite natural, a process requiring no great mental reservations on our part. Putting two numbers together, reducing a formula, or utilizing a pocket calculator or computer use the same functions. On the other hand, it is fundamentally unclear how concepts such as calculation, computation, and algorithm should be delimited—that is, what objects or processes these concepts can be applied to, and what they mean in a deeper sense. The lack of conceptual clarity fortunately does not mean that it is impossible to carry out computationally based research on nonlinear dynamic systems, chaos, artificial life, or neural networks. These contributions to a science of complex systems are not necessarily inhibited. On the contrary, new insights are constantly being produced, insights that can be used to understand self-organization, not only of life, but perhaps also of computation in nature. Still, the foundational questions are not yet resolved, and they might lurk behind some of the specific questions that we attempt to answer.

How does computation arise? This question can be addressed from several perspectives, according to one's temperament and professional background. Anthropocentrically, we could say that computation always demands a human being who interprets the manipulation of symbols. Even though a machine can compute, it is we who endow the input and output with meaning: input and output are "numbers," mathematical symbols, and not simply physical tokens or states in the machine or on the screen. Cognitive science may choose the perspective that thinking can be understood in computational terms, but here the very concepts of computation, physical symbol, etc., are simply presupposed. From where do the symbols come? They are the meaning that the human subject *intentionally* (by our ability to see something as signifying something else, or by our power of imagination) reads into the physical signs. But the mystery of where the original subjective meaning comes from in the first place remains unresolved.

Here we might emphasize that intentionality—in the sense of a goal-seeking directedness toward the outer world—has both a specific human aspect and a more zoological one. It is intimately connected with the body's sensing apparatus and orientation toward the environment. This attention, a requirement for moving around in our living environment, can also be found among other organisms with well-developed nervous systems. This aspect of human thinking should be connected to a biologically inspired concept of computation as something that derives from the bioneural network's symphony of informational interactions, something often known as the "subsymbolic computational soup." The concept of information used here does not presume that the smallest units of information are always those that have both reference and meaning: a genuine symbol (with reference and meaning) can be seen as a macroproperty of the entire collective interplay of the individual units and of the strengths of their connections (modified via learning) in the neural network.[95]

A symbol then becomes to a certain degree a virtual pattern with specific behavior, in the same way as the patterns in the Life game constituted virtual machines (e.g., for the production of gliders). This was because there was nothing in the game's "physics"—its cells and in the individual rules for updating their states—that specified anything about these machine's real physical existence. We may use cellular automata here as a model, perhaps even as a theory, for explaining the conditions for achieving computation in physical and biological systems. Let us take an example.

LIFE AND COMPUTATION ON THE EDGE OF CHAOS

The seven commandments of artificial life can be criticized for saying nothing in themselves about what distinguishes the living from the nonliving. They could just as well be seven commandments for an artificial physics. Two responses to this objection are possible. The first is to acknowledge that the commandments may be supplemented with several additional criteria in order to apply to the life we know (see fig. 2.3). The seven commandments can be used to distinguish research in a-life from traditional research in b-life, but not a-life from a-matter (understood as artificial physics). The second response would be to insist that it is precisely the very possibility of a continuous transition from artificial physics to artificial life that is the important point, since life itself must have emerged via self-organization of matter as autonomous units with the ability to store and process information. The notion of evolution rests on the notion that what is new is connected to the old and cannot be radically separated from it. However, just as biological systems cannot normally be half-alive, computation is typically viewed as an all-or-nothing phenomenon: either it is a matter of manipulation with symbols that stand for numbers according to specific rules for calculation, or it is not computation at all.

Langton has investigated the conditions by which computation may emerge in a physical system, or at least those states that operate as preconditions for computation. These preconditions include (1) the possibility to store information; (2) the possibility to transmit information; (3) the possibility for computation itself, that is, the modification of information when it interacts with other information (the latter is essentially what happens when we calculate 167 + 985 = 1152; we use 167 and 985 as the input, which we modify during the computation process to arrive at 1152). The physical system Langton investigates, however, is an artificial physics composed of cellular automata. Langton reformulates the problem of how computation emerged thusly: "Under what conditions will cellular automata support fundamental operations of information transfer, storage and modification?" From this, we can attempt to infer backwards and trace the conditions by which calculations could spontaneously arise in nature.

In a comprehensive study, Langton has simulated and analyzed dozens of one-dimensional cellular automata with various rules (CAs of the same type as Wolfram's, see fig. 4.6).[96] His investigations revealed that computation seems to be able to arise spontaneously in physical systems; namely, across the sharp boundary between automata rules that yield a highly ordered and very regular dynamic, and rules that yield a highly disordered, chaotic dynamics for a cellular automaton's development. This boundary, asserts Langton, corresponds in physical systems to the phase transition between a solid and liquid phase of matter. The phase transitions for H_2O are familiar to us in the melting of ice into water and the boiling of water into steam. Langton believes that life itself arose in proximity to a phase transition between solid and liquid. "Looking at a living cell, one finds phase transition phenomena everywhere," he says.[97] It thus appears that life and computation are closely related phenomena: life can be viewed as a form of emergent computation created at the very edge of chaos.

It is only after having investigated a wide range of cellular automata and their development that Langton reached the conclusion that in the narrow no-man's land between order and chaos, both static and dynamic propagating structures can be supported. The latter are reminiscent of the gliders in Conway's Life game. According to Langton, the Life game, which in contrast to Langton's own automata is two-dimensional, is situated within this very critical transition area. (This zone can be characterized objectively by a certain parameter that is found by analyzing the rule table by which a given CA evolves.) The static and propagating structures that appear in Langton's artificial physical system can thus form the basis for information storage and transmission, and the automata can also exhibit forms of "collision" between the patterns (fig. 5.2), which Langton interprets as modification of these signals.

This means that information becomes an important factor near the point of phase transition between periodic and chaotic behavior in cellular automata. In the very ordered

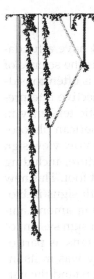

FIGURE 5.2. Langton's computer run of one of the automaton rules that gives dynamic development of propagating structures or "signals," which interact with themselves during the automaton's development. According to Langton, these correspond to information processing (the parameter mentioned in the text has the value 0.45). (From Langton 1990, see n. 96.)

area the patterns created by the cells are overdependent on each other and cannot enter into any computational tasks. In the chaotic area, the cells become too independent and are again unable to cooperate in any computation. Only when they approach the critical transitions can information be sent out over long distances without dying out. It permits the dance of long-distance behavioral covariation of patterns in the dynamic space, which is a necessary condition for computation: "Correlations in behavior imply a kind of common code, or protocol, by which changes of states in one cell can be recognized and understood by the other as a *meaningful signal*. With no correlations in behavior, there can be no common code with which to communicate information."[98] Langton does not directly assert that any of these automata's patterns in fact compute, only that they have the potential to do so. He has just given his interpretation of his study by suggesting that "computation as we know it is really just a special case of a more universal physical phenomenon" (p. 32).

Semiotic Questions

If Langton's interpretation is correct, it will have a fundamental importance for semiotics. Semiotics is the science of
signs and the creation of meaning. Semiotics is often considered to apply only to human signs and subjective interpretation. Under the influence of a new, realistic tendency in
semiotics, inspired by among others the American philosopher Charles Sanders Peirce (1839–1914), we now view sign
phenomena as occurring everywhere in nature, including
those domains where humans have never set foot. This new
field, called biosemiotics, concerns itself with signs in biological systems, ranging from communication among animals to the individual cell's genetic code as a sign system of
its own.[99] Peirce himself attempted to see signs as part of
natural evolution, but he died while biology was in its in-
fancy. Today we are better equipped to understand the natural history of signs. It is a history that dates back to the
origin of life, for living cells, in order to survive as complex
systems, had to possess a code or partial description of their
own structure, so that they could begin to collect descriptions of survival. It therefore arouses great interest that perhaps there is now beginning to emerge a physics-inspired
theory about the natural realization of computational phenomena, where computation can be understood as the manipulation of signs par excellence. Cognitive science, too, is
searching for theories about physical and biological conditions for the realization of computation. Langton's question, "How does the dynamic behavior in a physical system
become computation?" corresponds to the semiotician's
question: "How do information, sign, and meaning emerge
in nature?" We are beginning to identify associations between semiotics, artificial life, and more traditional research
in human thought processes.

There are many questions that Langton must answer, of
course. Is computation necessarily connected with phase
transitions? Why doesn't my computer sit there and cook

when I use it? The comparison between transitions in cellular automata from periodic to chaotic attractors and nature's phase-transition phenomena is exciting, but does it work? And if there suddenly appear signals—information as Langton confidently calls it—in a cellular automaton boiling on the edge of chaos, what is this information then *about*?

It is one thing to say that CAs and their rules can operate as the physics through which we can emulate virtual computers—that for example, a universal Turing machine can be implemented in the Life game. It is quite another thing, however, to assert that computation spontaneously appears from the borderlands of chaos in such a system. In one case we are speaking of a slightly peculiar but feasible way of building a calculator: namely, using von Neumann's method. We are speaking of a product of human design, and hence, a product that can be understood only as an intentional system. In the second case we are speaking of the spontaneous emergence of something new within a complex medium, without any sort of planning, and without requiring interpretation as a computation in order for us to understand it. We are not necessarily forced to use an intentional perspective on the signals that emanate out of the chaotic dynamics of a cellular automaton.

We thus have *two distinct concepts of computation* at work, and the question is whether in the next ten years we can get them to communicate.

1. We have a concept of computation as logical operations on symbols that can be defined purely syntactically, but which nevertheless signify specific numbers and mathematical operations (i.e., concepts with a semantic content) by which we intend to solve a specific mathematical problem. Even though the computation itself can occur mechanically, the physical signs first attain meaning during a cognitive interpretation as representative of a mathematical, logical, or other domain. The physical signs represent numbers. This is the *representational* concept of computation.

2. There is the concept of computation as processes that not only take place in a trivial sense in physical systems (or models of these) and are therefore dependent on their dynamics, but which are essentially inherent in these processes and demand no further interpretation or ascription of meaning. The computation here becomes part of the system's physical ecology, something that describes macrostates in the systems. This is the *physical*, or nonrepresentational concept of computation.

Semiotics, together with other disciplines, can contribute to connecting these two notions into a generalized concept of computation, even though this enterprise has to deal with fundamental questions of biology, physics, and mathematics.

AN ON/OFF UNIVERSE

Ecology analyzes the relations in nature between organisms and the inorganic environment. The ecology of computation can be viewed as the study of all those relations that deal with computation in physical, living, and artificial systems. What is needed in order to keep a computational system going? What "biotic and abiotic factors" are linked to computation? What limits for computation are set by its consumption of energy and matter? And so on. It involves many perspectives. We have known for a long time that there are physical limitations on the efficiency of computations. Computers develop heat, which they have to get rid of. This is primarily due to the fact that a computer consists of a mass of electronics and mechanics: rotating hard disks, read-write heads with step-motors, disk drives, transformers, wires, the electronic circuitry making up the semiconductor memory, a cathode-ray tube as the screen, a loudspeaker, etc. As in other machines these parts generate heat due to friction when they operate. However, the computational processes themselves—which in the computer demand the transfer and deletion of information in various storage areas and logical ports during the execution of computational processes—also require energy and generate heat. The heat generation fundamentally limits both energy efficiency and the speed with which computers can carry out computations, at least given the principles used for building today's computers.[100]

The ecology of computation entails considerations that can cause dizziness. Studying the physical limitations on computation entails acknowledging that computation, much as the measurement process in physics, is far from completely idealizable or meta-physical. Computational processes are subordinated to normal, natural laws. At the same time, these laws are described by mathematics and computations based on this mathematics. Hence, it is natural to conclude that the physical laws must have a form that permits them to be executed in time and space. The laws that describe our universe must be limited to algorithms that can in fact be executed using this universe's bricks and mortar. The universe does not permit us, for example, to build a computer with an infinitely large memory that could help us to separate pi from an arbitrary close-lying neighboring number. The physicist Rolf Landauer has pointed out this connection (fig. 5.3).[101] It sounds like a modern version of Hegel's dictum that "all that is real is rational, and all that is rational is real."

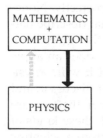

FIGURE 5.3. Information processing, shown above, depends on what the real universe permits in purely physical terms. These physical limitations on computation are shown with the right arrow. However, the laws of physics, according to Landauer, are "recipes for information processing" and are therefore dependent on executable algorithms; this is indicated by the left arrow (after Landauer 1987, see n. 101).

Some alifers state that they sometimes have a creeping feeling down their spines, as if they were being watched while they sit in front of their computer screens and design little universes with some artificial life. What now if they themselves belonged to such a little universe? They are tempted to look back over their shoulders for a Grand Designer who is "running" them on an even larger cellular automaton, some kind of *super*supercomputer.

We are all subordinated to fundamental laws of nature: the forms that nature has assumed as habits and in which it moves. In this sense, it is perhaps correct that nature calculates our next state for each nanosecond. If the calculations are deterministic, this idea corresponds to reviving "Laplace's demon," a being with infinite information about all states and an unlimited computational capacity, a being who could therefore calculate the state of the universe infinitely far into the future. The complexity of the world, however, keeps the demon from using a world formula for either shortcut calculations or predictions. (There are several additional reasons that make the demon's existence impossible.)

Galileo once said that the book of nature has been written in the language of mathematics. If this is true, God must have been a mathematician. Physicists often take for granted that the mathematical language of abstract forms is the most effective way of describing the universe. Yet it remains somewhat of a mystery. There is no philosophical or scientific guarantee that this artificial language, created by human beings themselves, depicts nature correctly. Nonetheless, the various differential equations that describe natural laws do keep track of things. What is it that gives these equations their strength? Why, for example, does an apple in free-fall act precisely the way we predict it will? Physicists do not know, and most of them will not discuss it; and philosophers of language simply state either that we are breaking the rules of language when we ask these kinds of awkward questions or that we are leaping from a scientific language game to a metaphysical one. This mystery should remain a private, irrational concern. The physicist Richard

Feynman freely discussed the mystery, however. He remarked that what is important about a physical law is not how clever we have been to find it, but how clever nature has been in respecting it.[102]

Edward Fredkin is a man who has taken the spine-tingling feeling seriously. He believes that the universe is one giant computer—a cellular automaton.[103] A given CA can imitate life, ice crystals, or galaxies. When the ice crystal grows by one small piece of ice, it corresponds to an extra cell being turned on in the automaton's lattice. Fredkin knows, of course, that such CAs are not ice crystals, but his idea is that cellular automata will depict reality more and more precisely if they are applied to ever smaller bits of matter. The rules then correspond to the movement of molecules, and at a smaller level to atoms, electrons, quarks, and whatever the entire family of elementary particles is called. At the most fundamental level, says Fredkin, the automaton will describe the physical world's forms of movement with perfect precision because at this level the universe is a cellular automaton in three dimensions. The world is cellular, a giant lattice consisting of fundamental grains or logical units, each of which computes at a completely local level their state in the next picosecond as a function of the states of the neighboring cells. The information processed at the basic level is the factory for the reality; it is the stuff that makes up the matter and energy of physics.

Fredkin's universe expresses a unique ontology (the study of being), and when he made his ideas public, he was not taken very seriously. As a self-educated computer scientist, Fredkin became wealthy running a computer company that specializes in image-processing. It was perhaps here that he learned how to view reality as bits of information. Fredkin has always had creative ideas, and today he is a professor at M.I.T. He has published practically nothing, but inspires others. As a result, many of his colleagues, among them Wolfram, are beginning to believe that his ideas are not so crazy after all. In Fredkin's universe, an electron is nothing more than a pattern of information, and an electron in a path is a pattern that is moving. Of course it does not really move, for movement is just as virtual as the gliders' movements in the Life game, or the movement of letters in a rooftop electronic-message board: the movement is illusory, created by the pattern of blinking bulbs. Fredkin's universe is a virtual universe.

This naturally gives us grounds for a degree of unease, and for many questions. Fredkin's automaton universe can perhaps explain how mathematical natural laws can move the world, for the laws are certainly implemented as rules in a larger automaton. The movement is, so to say, simulated, virtual. Yet if the universe is indeed a computer, and the physical natural laws its software, then we will always be cut off from knowing what the hardware is like. Turing certainly showed that computers are equivalent to universal machines, so that a program does not in any essential sense depend on being realized by a single specific kind of hardware. Since in Fredkin's universe we ourselves would already be virtual meta-entities in such a cosmic program, we could never know anything about the primary machine. We could hope that it never crashed, for we would go down with it.

It may well be that at the lowest level the universe *could* consist of discrete small grains of time-space units; it may well be that the universe is one giant, parallel, CA-like computer. However, it is doubtful whether Fredkin will ever find the great rule that can make a cellular automaton function like our own universe. One of the problems that the physicist Heinz Pagels has pointed to is that the cellular automatons can be created in classical machines, i.e., arrangements that can be totally described by classical physics, and that therefore have limited ability to simulate certain quantum-mechanic phenomena; it is these very phenomena that transcend a classical description. There is quantum-mechanic sand mucking up Fredkin's machine.

So even if the computational ecology of this new world science is an attractive paradise of the intellect's imagination, we should beware of having rational thought bewitched by metaphysical fantasies and unfounded games of language.

In the next chapter we shall consider the artificiality of a-life from a more critical point of view.

Chapter Six

THE BIOLOGY OF THE IMPOSSIBLE

SNAILS are just one example of the miracle of life. All of us are familiar with garden snails and their intricately swirled colored shells. After a rain, they crawl out onto roads and paths. The glistening forest snails are gorgeous creatures; some are the black gold of the forest, while the red variety are like flaming torches. There exist so many different species of snails that no one person knows them all. Snails are simply something special. The naked, shellless *Opisthobranchia* snails that live in the sea are among the earth's most exquisite creatures. But the bashful gray-brown edible Roman snails, *Helix pomatia*, also possess their own sublime beauty. And they taste so good if they are prepared properly.

FIGURE 6.1. Two edible snails meet.

It is, by the way, an opisthobranch, an *Aplysia*, which because of its giant, easily identifiable nerve cells has to donate its body—or should we say its foot—to an infinite number of neurophysiological experiments. The day someone finally maps out the total hardwire diagram of the *Aplysia's* nervous system will be the day when attempts are made to imitate it artificially. What is easily forgotten is that a single nerve cell carries out such complex information processing that at least ten neural networks will have to be created in order to simulate a single *Aplysia* cell, and this would still be at the level of a simulation.

Artificial snails? I don't believe we'll see them.

MAKING LIFE A BIT GREENER?

Artificial life can be viewed as the realization of a scientific dream of understanding and thereby controlling life processes at their very foundation. The dream has clear mythological forebears. One need only mention names like Frankenstein, the Golem, or Pygmalion, or think of the entire genre of science fiction. The dream shows itself today not only in biotechnology, but also in the use of the computer as an extension of the senses, or as some skeptics might say, as their amputation. That the computer itself has become the symbol of all of life's artificial aspects takes on special significance for the way we evaluate artificial life.

The computer is a tool and a medium. Like other tools, it increases our control over our surroundings and at the same time takes us away, alienates us, from nature. Now the kind of nature that is controlled by the computer is not so much the raw, primary nature as it is the mediating, second nature—the symbols, the language, and the rest of the human universe of signs. It is here that the computer shows itself as medium. It is a medium for communication, simulations, and games of every kind, including the scientific, logical, and mathematical games of language. Like other media, the computer also has an indeterminate social duality, increasing our contact possibilities and at the same time alienating us from each other. It is a tool for alienation but also for making us more cultivated, more civilized, and developing our common potentials. It is interwoven with other information technologies and with possibilities for misuse of power and control of human relations. And the computer has itself become a cultural symbol with many meanings that makes it almost impossible to face its products without an inborn prejudice. In the schism between the System's goal-oriented, calculating rationality and the probing, value-laden sensuality of common everyday life, the computer clearly belongs to the System.

Hence, we can also view computation-based artificial life as yet another attempt on the System's part, certainly at the theoretical level, to seize one of the free places otherwise retained for the green aspects of existence. Does artificial life make life any greener? There is not much evidence that it does. Artificial life is all too rigid, flat, without dimension, and calculable. Only the most far-out computer freaks can seriously talk about artificial life's human rights or the ethical problems in pulling the plug out of the machine. Artificial life on a computer has nothing to do with life, one may argue. On the contrary, it is an extra step in the direction of the sterilization of our environment and our possibilities to achieve an immediate, sensual relationship to the rest of nature.

Now there are certain important features that certainly do not make life greener, and that could make more valid claims on our keen indignation: gases generating the greenhouse effect, acid rain, our general squandering of resources to name just a few of the more life-threatening forms. I think our unease about artificial life is based on the fact that like the computer, it has become a symbol, a metaphor for something larger. It feels like the concept of life itself is being threatened, even though artificial life is formulated as a purely scientific project, and even though this dimension of the life concept lies outside the scientific domain.

This threat has appeared in much stronger form in connection with the debate on the entire medical technology of birth and death. One might think that this debate is something completely different. In fact it is not so different at all. Both a-life and the discussion about death-defying medical technology revolve around the idea that life as a biological phenomenon can be made the object of definitive, objective description, a thought that becomes repulsive to us, if not threatening. The System does not respect the boundaries of the human world, but seeks to carry out its idea of calculated objectivity at the expense of this world.[104]

DEMAND THE RETURN OF DEATH!

When living biologists compare artificial life with real life, there is one difference that quickly appears: artificial life is life without death. Or in any case, a life whose death is very clinical, artificial, and digital, a momentary change of state from on to off, with no resemblance to biological death. In the world of biology, death is a process that itself involves new life. Life requires death. Computer organisms do not die at all in a biological sense; they simply cease to exist. But real death is part of the living game. Here, too, there is a problem with the relationship between organism and environment, because death in an organism is a process where its delimitation vis-à-vis the environment is slowly erased, and where the substrata, nutritional salts, water, and other materials of which it consists are released into the environment and can recirculate in the ecosystem itself. This could perhaps be represented in expanded a-life models, but it would ultimately demand modeling every single atom or molecule, to the point that there would be no reason to have a model at all. Life is also death, dissolution, dung, and earth. The problem more than hints at the fact that artificial life is located within that domain of human construction we call language, and is beyond life and death in a biological sense.

The digital and somewhat clinical aspects of computer organisms will enable biologists to demand the return of death to biology. Life as form is conditioned by a concrete material, a slimy substance that can be dissolved and reformed into new creations. Artificial life is but a repetition of an ancient occidental mania for repressing all that is connected with the body and the senses. As the social critic Theodore Rozak asked:

> Since we were children, what have we been taught to regard as the quintessential image of loathing and disgust? What is it our horror literature and science fiction haul in whenever they seek to make our skin crawl? Anything alive, mindless, and gooey ... anything sloppy, slobbering, liquescent, smelly, slimy, gurgling, putrescent, mushy, grubby ... things amoeboid or fungoid that stick and cling, that creep and seep and grow ... things that have the feel of spit or shit, snot or piss, sweat or pus or blood ... In a word, anything *organic*, and as messy as birth, sex, death, and decay. We cringe from anything as oozy as the inside of our body and look for security to what is clinically tidy, hard-edged, dry, rigidly solid, odorless, aseptic, durable. In another word, anything lifeless—as lifeless and gleamingly sterile as the glass and aluminum, stainless steel and plastic of these high-rise architecture and its interiors that now fill the urban-industrial world.[105]

ARTIFICIAL RESEARCH

How has the message of artificial life been received by the biological community? A-life ought to be of major interest if it can lead to new biological generalizations, solve the riddles of the creation of form, improve our understanding of the evolution of complex living systems, and introduce better methods to achieve a coherent theory for synthetic construction of organisms. Is a-life research a decisive breakthrough in theoretical biology, or should it be considered a curiosity?

The first conference on artificial life was held in 1987. The interdisciplinary gathering around a research program for a computation-based exploration of synthetic living systems is still new and has led to only a few reactions from biologists not involved in one of the a-life projects. However, this is just a question of time. In the 1950s, when logicians, computer scientists, philosophers, psychologists, and linguists began to collaborate in the field of artificial intelligence (AI), heated debate arose. Many psychologists considered the project—and continue to regard it—with great suspicion. The history of AI is therefore also a continuing discussion between enthusiasts who proclaim rapid progress and critics who argue that the project is fundamentally impossible. Overlooked is the large middle group within the discipline who methodically and pragmatically carry out work on what science historian Thomas Kuhn has called "puzzle solving," the tedious research work that seeks to solve some of the small problems that in fact can be solved within the framework of methods and perspectives possessed by a given discipline.

The criticism that biologists bring to artificial life resembles to a certain extent the criticism of AI, but the debate on AI has also changed. Artificial life itself already represents a contribution to the debate on artificial intelligence; namely, that it ought to be constructed "from below" and start out from simple biological or bio-logical principles. Some biologists will surely feel tempted to believe that from a natural-science point of view, the exploration of artificial life is not much more than artificial research. Whether the criticism of artificial life becomes just as livid will depend on the results that the alifers can generate during the next five or ten years, and on the funds allocated to the research program.

Let us examine several of the objections and open questions posed to a-life research by biologists and philosophers.[106] Some of these objections contradict each other. It is premature to provide an overall evaluation of the research program and its basic ideas; rather, my intention in considering these criticisms is to shed light on some of the central problems in artificial life.

Neo-Behaviorism?

There exist a range of arguments against artificial life, not all equally strong. We can take as our point of departure the seven commandments (cf. chapter one), formulated here as the anti-theses of a critic.

1. The Biology of the Impossible

Artificial life is not the biology of the possible (first commandment). It may be computer science and mathematics, but as biology it is impossible. The reason is simple: biology must be founded upon the exploration of life as it actually appears here on this planet, because one of the most important features of life is that it is *unique*. We can certainly not give a definition of life, but we can give an incomplete list of certain properties associated with living processes. Ernst Mayr, one of the founders of the "modern synthesis" or neo-Darwinism in biology, has composed such a list, and we can simply mention the point that Mayr calls uniqueness.[107] The chemically unique feature of life consists in the fact that in no other place in nature do we find the macromolecules that we find in living cells: nucleic acids, proteins of various kinds, membrane lipids, etc. To believe that we can abstract this away and at the same time maintain that we are still investigating something living is an illusion. Besides, life is biologically unique because we cannot find two organisms that are identical in every respect; life exhibits an enormous variability that in a radical sense distinguishes organisms from objects in the physical or mathematical sciences. There is nothing there that resembles the drastic changes from birth to death that characterize living beings. On this point, Mayr believes that biology—which is intrinsically related to life's historicity—radically distinguishes itself from the concerns of the physical-chemical disciplines.

One of the problems with this objection is that it builds upon an outdated view of physics and mathematics as identical to, respectively, mechanics and logic. Many of the features Mayr cites as characteristic for life (e.g., "indeterminism," whereby he means the emergence of new, unpredictable qualities at a higher level) are now preoccupations of physicists studying nonequilibrium thermodynamics, chaos theory, and artificial life. By emphasizing chemical uniqueness, a virtue is made out of what the alifers call carbon chauvinism. Rather than disproving the idea of life in other media, carbon chauvinism simply precludes it out of hand. One could instead discuss the *relevance* of sacrificing scientific resources—people, time, money—to investigate life in alien media while we still know so little about so many areas of life here on earth.

2. The Asynthetic Objection

The attempt has been made to appeal to a holistic method by asserting that a-life can synthesize life (the second commandment), while traditional research in b-life can only analyze and destroy life. However, this assertion, say critics, is in many ways misleading. First, biology today contains not only analytical fragments of knowledge, but in fact also contains broader theorizing about evolution, cell biology, and the entire ecology of a wide range of related insights. Second, one could say that a complex whole such as an organism cannot be "synthesized" in any external way; it is precisely its organic autonomy, or what Maturana calls autopoiesis, that defines it as a living entity. It is not the extent to which it eventually could be studied in a more or less holistic fashion.

3. The Objection of Unreal Life

The so-called computer organisms discussed in various examples and models are not and never can be genuine life (contrary to the third commandment). Alifers certainly do not assert, for example, that the individual machine codes in VENUS or the individual boids in Reynolds's simulation of a bird flock are authentic; the assertion, rather, is that the behavior that emerges out of the interactions of the individual parts in the model is genuine behavior or genuine evolution; it is just as authentic as real life. However, this is only a purely external consideration, a neobehaviorism that says that when something resembles what it represents, i.e., when its input/output functions are the same as the input/ output functions of a given organism and a black-box model of the organism, that the black box must therefore be alive. This is a very superficial criterion for life, no better than animated mechanical animals constructed of screws and gears.

Turing once proposed as a thought experiment a test that could determine whether a machine is intelligent. A subject is allowed to communicate with a machine (by letter, for example), without knowing whether the recipient of the message is a machine or another person. If, from the responses to his questions, the subject cannot determine whether they come from the machine or from another human being, then the machine has passed the test. The comprehensive discussion of the Turing test, which derived from the critique of AI, has made it clear, however, that such a pure behavioral test is not satisfactory. For example, it is relatively easy to construct programs that answer questions in ways that resemble human responses, without the program at all representing the least linguistic understanding or especially intelligent behavior. Similarly, a purely behavioral test for artificial life is not especially informative. Not much is required to present a program according to the principles of a-life (a bottom-up specification of the program and parallel computing units, e.g., as in the Life game), and achieve interesting behavior, which is certainly complex, but which contains no more biology than a game of billiards.

4. All Life Is Matter

This half-truth can be just as valid as the idea that all life is form (the fourth commandment). In one sense, biologists acknowledge the correctness of Langton and Farmer's observation, that the particular type of matter is not especially essential for biological form. The molecules in our own bodies, from skin to joints, are constantly being replaced, so it cannot be these individual molecules as such that define the form. Nonetheless, embryologists and molecular biologists both point out that the form of, for example, a fruit fly's body is determined by genes that code for morphogenetic substances, and that one needs to know the specific structure of these substances in order to understand the mechanism of the generation of form. These mechanisms are as yet virtually unknown. Also, before we can explain that a given biochemical element of the body can have a specific biological function, it is not sufficient to know only the functions in detail. We must also know the specific structure of the given substance. Nor can we understand a living cell in total abstraction from its material components, atoms and molecules. It is this feature on which molecular biology bases its success. It is also here that von Neumann, as we have mentioned, expressed his reservations about pure formalization of something as organic as self-reproduction.

The problem of relating the formal and material descriptions of living beings has been succinctly summarized by the Danish biochemist Jesper Hoffmeyer, who states that life contains both a digital self-description (in genetic form) and analogical aspects that are not linguistic, but that interact with the physical environment. The American physicist Howard Pattee has also used a variant of the digital/analog metaphor to describe a living cell; Pattee discusses the complementarity between the cell's linguistic (the DNA code, the self-description) and dynamic aspects (the physical/chemical work executed by the cell).

Nevertheless, in its own back-handed way, molecular biology confirms the thesis that life is indeed form: if we seek to understand the material basis for life processes via biochemistry and molecular biology, we must nevertheless concentrate on form—but simply at a subordinate level. What at one level (e.g., the cell) appears as the "material," which supports the cell's specific forms, manifests itself at the next lower level as new forms (the macromolecules' specific forms, which determine the enzymes' affinity for various reactions). These forms have themselves a material foundation at an even more basic level, and so on. Materiality ultimately dissolves into several organizational forms at different levels. This idea is hardly revolutionary: for Aristotle, form and material were complementary concepts. The material (wood, for instance), which is formed into a specific object (a chair) has its own form (woodness). One can therefore always ask of what more basic material (carbon) this form (woodness) has as its foundation, and what form this foundation has. Unlike Aristotle, however, we do not know whether we will ever reach down to a prime material, which is completely without form.

5. *Life Breaks Rules*

Considering them as principles for mathematical models of living systems, one can hardly make any biological criticisms of the three final commandments, (bottom-up construction, parallel processing, allowance of emergence). Here the critical argument is as follows: if artificial life is to be constructed, it is reasonable to recall these principles, but for other reasons it is doubtful whether trying to implement life purely digitally can result in something termed life.

A biologist fond of living spontaneity might interject that life in cellular automata (regardless of whether they are quiet life, dirty smoking trains, or smooth gliders) always follows the same simple, deterministic rules. Genuine biological life—what alifers would like to degrade to b-life, perhaps denigrating biologists as b-researchers—is characterized by its capacity to *break* rules. Life is certainly a game, but it is not played according to fixed rules. The rules themselves can evolve, and during evolution old rules can break down. The models with which alifers work are precisely that: only models. They do not capture life's anarchical tendency to break rules, especially the rules that biologists themselves have attempted to construct as general biological laws. Biology belongs to one of the surprising sciences, where each rule must always be supplemented with several exceptions (except this rule, of course).

6. *The Objection of Emergence*

Emergence is many things. It is possible that a-life is emergent (the seventh commandment), but it is unclear what exactly this is supposed to mean. Life as a biological whole is an open, uncompleted process, where new properties arise, both in an evolutionary sense with the creation of new species, and during the evolution of the individual, when the genotype (to say it too simply) unfolds into the phenotype. New species cannot be "derived" in any deductive way out of old ones, and phenotypic properties are not explicitly coded in the genotype. Hence, emergence is a fact of biological systems. In purely physical systems as well, new properties emerge that were not present previously, or properties emerge at a higher level than that of the component individual parts. Emergence is an unclear and inadequate criterion for the authenticity of artificial life.

The remainder of this chapter will discuss these critiques in more detail.

IN THE EYE OF THE BEHOLDER?

The assertions that the logical form of an organism can be separated from its material foundation and that the organism's ability to live and reproduce itself should be a property of its form alone both appear to go against the intuition of a molecular biologist. The empirically rich concept of the cell as biological unit implies that form and matter are not separate. And if the thesis of life as pure form does not apply to the cell as the fundamental biological unit, its validity in other parts of biology must certainly be questionable.

It is often imprecise what it is that is denied when something as metaphysical as artificial life is on the agenda. Let us try to clear up this problem by presenting an argument against the strong version of a-life, i.e., the version of the research program that stresses the idea that life is a medium-independent phenomenon and asserts that when, by program-governed computation, emergent lifelike properties appear, we are in fact speaking of a *realization* of new forms of life (in another medium), and not simply a *simulation* of life.[108]

It is first necessary to state what the counterargument does not say. It does not deny that in other physical media (e.g., on other planets), there could arise phenomena that we would term, by all criteria, living. It is conceivable that life is not bound to a specific carbon-chain chemistry, but it is probable that the ways in which life will self-organize in alternative physical media—i.e., the concrete embodiment it achieves, the species that will appear, and the types of processes it contains—will depend on the specific physics that characterize the given medium. That is, the independence of life-forms from the medium (in the strong sense) is denied, but the possible realization of life in other physical media is not excluded. These other media, which may also be characterized by physical parameters such as gravitation and the like, will simply place other limitations on the life-forms that might evolve. A smaller force of gravity, for example, will allow for larger animals.

Second, the counterargument does not deny that it is always possible to simulate self-reproduction and other aspects of living systems, just as one can simulate mathematical models of anything, from the weather to the national economy. One must, of course, always ask whether a given simulation or model captures some essential features of the system. Are von Neumann's and Langton's models of self-reproduction, which are based on a formalization of the biological problem, satisfactory? If we wish them to simulate certain mechanisms of the cell's self-reproduction, how do we then choose the interpretation of the formal model that best agrees with our other knowledge of biological cells? However, it is the assertion that the model achieves full and genuine life (or self-reproduction, etc.) simply due to its being constructed according to some specific computational principles that is denied.

The counterargument itself is simple and builds on certain relatively uncontroversial premises:[109]

1. Computer programs are based on manipulation of distinct symbols according to formal rules, i.e., they are built upon a discrete, digital syntax. This also applies to cellular automata, for they are formally equivalent to programs that can be run on normal sequential machines.
2. The existence of biological cells presupposes the process whereby cells carry out self-reproduction. This process comprises, but is not identical with, the copying of information. The cell's division is a material process, and the division includes the duplication of the genetic material that occurs via DNA replication (copying of genetic information). Life as a coherent phenomenon requires self-reproduction.
3. Syntax or "digitalness" in itself is neither a generic characteristic nor sufficient to constitute self-reproduction.

These three premises lead directly to conclusion (a): Programs are not in themselves adequate to achieve self-reproduction. A program cannot reproduce itself. (That it can copy itself is another, similarly doubtful assertion). This conclusion does not say that humans will not one day be able to build an artificial system that is alive, and that can carry out self-reproduction. It does say that it cannot occur only on the basis of computational manipulation using formal symbols (which is hardly surprising). Whether it occurs serially, as in classical computers, or in parallel form is of secondary importance, for in purely computational terms these different architectures are formally equivalent.

A fourth simple assumption is that the biochemical network of different metabolic processes, together with the enclosing envelope formed by the biological membranes, are sufficient for constituting a functional self-reproducing unit (the cell), and thereby function as a simple organism.

From here follows the somewhat trivial conclusion (b): Every other system that can self-reproduce must possess causal powers that at least correspond to those of the cell (physical/chemical interactions similar to the cell's are required). This will also apply to an artificial system. We thus reach conclusion (c): An artificially created, living, and self-reproducing system cannot exist simply as a result of the execution of a formal program. From this we derive conclusion (d): Cells do not self-reproduce or achieve other essential biological properties simply by the execution of a computer program.

A cell's activity involves something other than executing a program, while the latter is what essentially happens when Langton runs his cellular automaton. Langton's loop is certainly an example of parallel computation and purely local interactions in contrast to the programs that simulate intelligence according to "top-down" methods, as AI does. However, this remains computation: But did we not see precisely new patterns emerge and divide again? Were we not surprised about how lifelike they were? Yes, but here we must distinguish between the "emergence" that the simulations create for us when we observe them and the phenomenon of emergence itself. That the patterns created by cellular automata during their computational development appear to us as real is not the same as saying that they realize processes that have the same properties as those of nature, on which they have been modeled. The reality of a glider or a loop as a pattern remains a virtual reality. It is perhaps worth emphasizing, with all respect to Langton's models, that we have yet to build a self-reproducing machine.

The later critique applies not only to self-reproduction but to all forms of computationally based emergence. The system theorist Peter Cariani has tried to explain why this is so.

EMERGENCE'S MANY FORMS

In April 1990, three questions appeared on the computer screens of those alifers linked to the A-Life Electronic Mail Network. The message was from Cariani, and his questions were as follows: What distinguishes an emergent computation from a nonemergent one? What distinguishes an artificial-life simulation from other kinds of simulations? Do these distinctions depend on the formal behavior of the computations that are executed, or on the interpretations given by the external observer/programmer?

The questions were accompanied by a laconic remark that unless we clarify the definition of life and a method for distinguishing living organization and behavior from the nonliving, artificial life will remain a specialized branch of computer programming and not an independent field of research with profound implications for the rest of science.

The arguments Cariani has provided for his views are informative, not only for our understanding of "living machines," but also "thinking machines," "neural networks," and the like.[110] The core of his argument is that both humans and other living organisms stand in a relation to their surroundings that is not only computationally based, but based on measurement (i.e., sensing of the environment) and operative activity (i.e., active execution of controlling behavior). While computation can be abstracted from the physical substratum, measurement cannot. Computations, inasmuch as we can describe them as deterministic state transitions (and this applies by definition to computational phenomena), permit us to abstract from the material substratum and thereby allow a certain degree of Platonism. Measurements, in contrast, are always nondeterministic. We never have full control over the relationship between what we measure and the instrument we ourselves use to measure: full control would require complete information, and that would make the act of measurement itself meaningless. This aspect of the measurement situation was remarked upon by Niels Bohr and is part of the "Copenhagen interpretation" of quantum mechanics. Measurements are nondeterministic and relative in relation to an observer and that which is observed. In contrast to the purely mathematical description, we cannot abstract the substratum away (hence, an Aristotelian form/matter notion goes better with a "measurement philosophy"). Through measurement, sensing, and activity, we stand in an open and pragmatic relationship to the world.

This pragmatic relationship characteristic to genuine life is fundamentally different than computer life, whose environment is simulated. Computational forms of life do not stand in the same open, nonspecified relation to its environment because the environment, too, is of a purely computational nature: digital and determined. It is this basic insight that Cariani has refined in the following argument.

One can sketch out the sign relations internally in an organism or an artificial arrangement of the relationship between the organism and the environment in the following way (fig. 6.2): there are, according to Cariani, three kinds of processes at work:

1. *Nonsymbolic interactions*, which are "pure analog," not measured, and take place in the physical environment (neither the "input" nor "output" processes involve the use of symbols)
2. *Computations*, which are purely formal, rule-based manipulations with symbols (both input and output are symbolic)
3. *Measurements* and *control operations*, which relate the symbols in the formal part of the system to the external world as such. In measurement, input is nonsymbolic and output symbolic; in a control operation it is reverse.

One can also say that the computations are logically necessary, conventionally determined *syntactic* operations. Measurement and operative activity or behavior are, in contrast, empirically contingent, materially controlled, *semantic* operations, because they endow a kind of biological meaning to the organism. They imbue the computations with meaning for the given device or organism (which contributes to survival and performance, if the semantics are suitable). The choice of suitable measurements, control operations, and computations, which is decisive for how much of the device's model of the world meets the requirements for the device's function, ultimately constitutes the goal-related *pragmatic* operation.

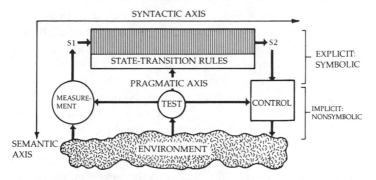

FIGURE 6.2. The basic sign functions in an organism or model relations in a machine, e.g., a robot with sensors and a control program. The computation is purely syntactic. Measurement and control concern the relationship to the environment and are therefore semantic. By measurement and sensing, a nonsymbolic input is transformed into symbolic output. Conversely, the symbolic results of the computations are converted into nonsymbolic, control-type interactions with the environment. In evolution or in human design situations, the previous values of measurement, computation, and behavior operations are tested and changed. This designates the pragmatic aspect. (From Cariani 1992, see n. 110.)

This model of the syntactic, semantic, and pragmatic relations for a given physical system makes it possible to provide certain precise conditions for emergence: Cariani calls it the Emergence-Relative-to-a-Model perspective, which he opposes to the alifers' less precise concept of computational emergence and to the thermodynamic emergence concept used by certain physicists. For Cariani, emergence reveals itself as the behavioral deviation by the physical system from the observer's model of it. If we as observers look at a physical system that radically changes its inner structure, and thereby its behavior, we need to alter our own model of it in order to be able to observe the system's behavior (and eventually make predictions about it). If we can continue to track the system with the model we already have, then nothing essentially new has arisen in the system, and we will be unable to make any conclusions regarding emergence.

A system is *syntactically emergent* if the pattern of syntactic state transitions is changed and new computational transitions are created. For example, one can train a neural network to change its computational state transitions in relation to the external environment. If the observer chooses a model of the network congruent with the machine's computational states, the pattern of computational transitions will change between each round of training, and the network's behavior will appear emergent.[111] A system is *semantically emergent* if new measurements or control operations are created (new in relation to the observer's model). If, for example, a measurement is emergent, it corresponds to the organism having acquired, in the observer's eyes, the ability to sense qualities in the world that were not previously represented in the observer's model of the organism's sensing. Cariani believes that cybernetic arrangements with semantic emergence are largely unknown today, aside from those which nature herself has invented.

The most widely known computationally based systems are digital computers with programs that are limited to prespecified rules and elementary computational states. Our observational frame of reference contains the potential machine states. Even though there may be many of them, our model of the machine is entirely state determined. As long as there is no hidden input that falls outside our model, the machine will continue to behave in exactly the same way, given the same initial state. If we operate the machine with the same input data, no new state transitions will emerge. It is certainly designed to function independently of variations in temperature and similar disturbances from the environment. This is also true for simulations, which have as their state space a subset of the total machine states. As long as we observe the total program, the total input, and the total amount of permitted states, the simulation will yield for us exactly the same behavior every time, and there will be no emergence in relation to an observer. The observer's model of the simulation will not deviate from the digital computer's running of the simulation.

This is also true, of course, for a-life simulations or for systems like Langton's emergent computations; these systems are located at the edge of chaos in an artificial physics. All computer simulations can be described as finite-state automata: formal systems for the manipulation of symbols. As observers, we can in principle always find a frame of reference (a model), which makes our simulation nonemergent. This is because cellular automata are equivalent to universal Turing machines, or finite-state automata. If we choose to let the variables we observe to be coextensive with precisely the stable computational machine states in the automaton being implemented by the simulation, we will always see it as a nonemergent state-determined system.

Talk of open-ended and indeterminate evolution and the emergence of new properties at a higher level in a-life models, or talk of attempts to increase the size of the simulation, or to simulate random and chaotic processes, or to represent the relationship between genotypes and phenotypes, and similar such strategies, cannot alter, according to Cariani, the basic replicability of these simulations. To speak of patterns that appear out of chaos or something similar—once they were not there, now they are—is also somewhat misleading. To investigate whether something is emergent, one cannot suddenly change the frame of reference from talking about microstates like pixels on a screen to gliders, wave signals, or other macrostates.

The important aspect of emergent events, artificially provoked by the simulation of life, lies not in the simulations themselves, but in the fact that they change the way we think about the world. Rather than being emergent arrangements in themselves, Cariani sees simulations as being able to set off emergent processes in our own minds.

Cariani's skepticism about the assertions of artificial life here and now does not, therefore, rest on self-reproduction being essential to life, something not yet imitated in the form of autonomous self-reproducing machines. Nor is he concerned that life is a slimy affair, which essentially demands precisely that slimy biochemistry with which we are familiar. Rather, life is slimy in another way. It stands in a permanent nondeterministic pragmatic relation to its environment. This relationship is computationally unpleasant, because we find ourselves unable, in a purely syntactically formal fashion, to capture the measuring behavior and evolution in the biological processes, the behavioral adjustments, and all the flexibility necessitated by an unstable environment and by the impossibility—both for us as observers and for the organisms themselves—of fully controlling the relationship to the environment. In principle, it is this indeterminateness that in critical seconds makes it possible for an animal to jump for its life.

RE-KANTING THE RULES

Perhaps we ought to revive Kant (or his medium-independent ideas) and the distinction he used in another context.[112] This is the distinction between (a) systems that *follow* a rule, i.e., consult a representation, some symbols, and make computations based upon them, where the nature of this representation guides the activity; versus (b) systems that operate *in accordance with* a rule, i.e., where the system behaves as if it had consulted a rule, but where the rule-like nature of the system, rather than being due to any inventory of representations, is instead caused by purely natural law–like causality. Even though my digestion is normally quite regular, my stomach is no computer, which for each new piece of meat searches its memory for a meat-digestion program. This does not prevent me from describing my digestive system as if it were a computer, however.

An apple falls to the earth. The path of its fall in the first t seconds, in conformity with Newton's laws, is equal to half the product of the force of gravitation and the square of t. The apple follows no rule. It moves in accordance with the law of gravitation. We cannot arrive at a computational theory about the falling without trivializing the concept of computation (as if it were the apple that computed its fall based on the earth's gravitational pull). The same can apply to my digestive system and to many other biological processes. This argument is important, for it also implies a rejection of the computational approach to physics in the strong sense, whereby people like Fredkin view the universe as one giant cellular automaton: the physical processes in the universe are updated not by consulting a table of state transitions or similar representations of natural law. We do not normally think in this fashion.

However, we cannot get very far with purely normal thoughts. Artificial life invites us to radically rethink our ideas: we risk finding out that our theories rest upon a foundation of quicksand, but we gain the opportunity to extend the insights that our snail's shell affords us. We must not be afraid to get a little slime on our fingers.

Chapter Seven

SIMULATING LIFE:
POSTMODERN SCIENCE

SITTING in the audience at a conference on artificial life, we become slightly confused when we see projected on the screen the many possible and impossible ecologies of computer organisms generated by computer simulations. After five hours of presentations of artificial universes, one gets the urge to grab for a real piece of fruit and eat a real steak in order to satisfy one's quite unsimulated hunger. It is like being caught in an immaterial universe of possible worlds, without a map that can lead the way out of the maze and back into a cafeteria in the physical world.

Let us examine some of the questions that researchers in artificial life are confronted with when they present their models and virtual organisms to a flock of bewildered biologists: How do you figure out whether the simulation is correct? What are the criteria for valid models chosen by the real parameters? What does a-life reveal about the evolution of real plants and animals?

The reference to reality may be disappearing insofar as one accepts a-life as a contribution to biology or other parts of research in complex systems. The reference to reality consists in the fact that experimental natural science creates a context where the symbolic language of theoretical concepts—via measurements and observations—is connected to nature's own signs, to the reality with which the theories deal. The realistic view of the theory is not based on a philosophical doctrine. Rather, it is a deep intuition that the world is there, and that we can describe it in language that enables us to depict the world. The realistic intuition says that theories are not only human constructs but are based on reality, and that steps toward scientific progress can continually deepen our understanding of nature. This conception is viewed by most natural scientists as a basic element in research. Critical philosophers have compared it with the quite naive notion of language as "the mirror of nature," but realism is also contained in far less naive incarnations.

Just as a skilled butcher slices up a steer—by letting the knife follow the natural openings and holes in the animal's body, without chopping, without cutting through joints and bone, but with a few strong swoops—the researcher must locate the categories and associations that allow the distinction that corresponds to nature's own types, and the simplest possible explanation that does not do violence to the phenomenon that must be explained.

The realistic intuition can be described in many ways, but let us simply choose the entirely schematic one: science draws a map of reality, creating a copy, a depiction, that says something about the world as it is, not just as it immediately appears, but as it *essentially* is. Biology must investigate and depict the mechanisms and laws that comprise the basis for the often chaotic and flickering confusion of behaviors in the biological sphere (fig. 7.1).

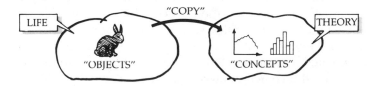

FIGURE 7.1. The traditional view of biology (as truth-seeking science) and its object (life as it is). Theories about living things are a kind of map that copy central aspects of the landscape. The map is not the landscape, but if the theory is adequate, it serves as an effective aid to our understanding of the object.

To call this view naive is simply to imply that it expresses the natural attitude. It can then subsequently be modified by asking more epistemological, critical questions, e.g., whether data is not always determined by the theory already being used; then the information copied into the theoretical field is itself colored by the questions asked of nature, etc. Most people will admit that things are so, and intuitive realism is therefore modified in the philosophical incarnations of realism, pragmatism, instrumentalism, etc. Nevertheless, we tend to maintain the view that the theory is a depiction of reality, albeit a slightly veiled picture, approximated, distorted, fractal (see p. 13 above), or in other ways incomplete.

Our unease with artificial life is due to an uncertainty about the question of whether it really is nature's own games that are depicted in the various simulations, or whether it is the autonomous games of an independent mathematical sphere that are given life and are achieved without any reference to reality itself. The naturalism of natural science is perceived to be threatened. However, it is not only in the scientific field that developments in computer and information technology have threatened naturalism.

A POSTMODERN SCIENCE?

The research program for artificial life can be seen as part of the much-discussed postmodern condition within philosophy, literature, art, politics, social sciences, and in social life generally, as it unfolds in our industrialized, information-dominated, media-oriented society. Artificial life indirectly confirms the contemporary diagnosis set out by philosophers, aestheticians, and sociologists such as François Lyotard, Mario Perniola, and Jean Baudrillard.[113]

In all areas of life we encounter a kind of distancing from the real and the material—a derealization. Communication technologies replace direct contact with the environment. The mass media create their own world, a hyperreality where the gap between image and reality is erased because everything is mediated, filtered through commentators, news bureaus, images, standardized analyses, talk shows, and events that are chosen and staged for the occasion. The media coverage of an event becomes the event's simulation, a complicated game that for all parties becomes more real than the reality itself. We never gain access to the raw reality, because reality itself is spun into the entire information society's web of images, messages, signs, and mediations. Substance disappears.

Even the sciences dematerialize matter, when they trace the world's smallest basic unit back to immaterial states of energy that themselves are described with mathematical formulas. High-tech measuring apparatuses lose their character as innocent information channels to the world and become filters that screen us from the world and bring us messages about conceptually difficult realities that can only appear as objects on a data screen.

Several developments within art can be interpreted as indicating this movement away from reality as object toward immaterial forms as something to be produced, or rather, simulated. Art has long transcended the classical idea of naturalistic depiction, where the picture was simply a faithful copy of the real (art as *mimesis*). Today's postmodern art and architecture also transcend the modern idea of the creating artistic subject, who in a sovereign fashion generates originals by natural creativity (art as *poiesis*). Instead, art becomes a simulation where copies enter into a combination of significations that are not actually new, but which represent small games that can be transmitted onward in a timeless infinity of circulating signs.

The picture becomes a *simulacrum*, a picture without precedent object, model, or ideal—a picture that meets no requirement for truth, individuality, or utopian hope. These metaphorical demands on the image are dissolved in a series of rituals that organizes the continued simulation of art in the universal media of mass society. The building becomes a mosaic of fragmentary stylistic references, a purely simulated aesthetic that masks every demand for faithfulness and suitability of form in a revolt against functionalist architecture.

Simulacrum is thus the concept for that which has broken

with any reference to a primary reality, and which is alienated from any original meaning or substance of material character. As image, the simulacrum is neither copy nor original. It has put behind it any precedent, understood as representation of something other than itself. As a media stream, the simulacrum is the place where personal experience and electronic news and multimedia experience are mixed together and become one. The visual media present and dissolve reality in their own image. The same tendency toward dissolution of difference between copy and original can be deciphered in the new sciences based upon computational technology.

It is characteristic of the simulations of life produced by alifers and cellular-automata specialists that these simulations do not derive from any natural domain, let alone an experimental physical system. The reference to substance, if not lost, is then at least left to others. Instead, the models help to produce a fictive reality, where the distinction between copy and original, description and reality, is meaningless. The model no longer seeks to legitimate itself with any requirement for truth or accuracy. It creates a simulacrum, its own universe, where the criteria for computational sophistication replace truth, and only have meaning within the artificial reality itself.

The relation to the objects themselves no longer constitutes a point of departure for efforts to achieve a cohesive understanding of a nature independent of us, an effort that modern natural science has inherited from natural philosophy and mythology. Science loses its mimetic function. Instead, the capacity to simulate entails increased production of new, immaterial realities whose complexity is apparently not surpassed by nature herself. The exploration of this universe certainly lies in the extension of the traditional empirical natural sciences. At the same time, however, it has freed itself from any constraints of a given material substance with a limited set of natural laws. Artificial life and artificial physics make it possible for the researcher to be the cocreator of those natural laws or rules of the game he or she wishes to investigate: it now becomes possible and desirable to simulate universes that are governed by other laws—another law of gravity, another relation between strong and weak nuclear forces, another speed of light, etc. The researcher creates a *virtual nature*, which at the deepest level makes possible the determination of the limits for the possible under other conditions. It is, of course, no real space in any empirical sense that is being researched, and the models will still have a quite local character. However, they will be able to simulate realities at every level, and thereby lead to answers to questions like: "Could our universe have had another structure?" and "Could we have asked this question if the known universe did not have precisely these natural constants, this mass, and these asymmetries between energy and matter as its fundamental characteristics?"[114]

From this perspective, artificial life must be seen as a sign of the emergence of a new set of postmodern sciences, postmodern because they have renounced or strongly downgraded the challenge of providing us with a truthful image of one real world, and instead have taken on the mission of exploring the possibilities and impossibilities of virtual worlds. It is a case of *modal* sciences, passing freely between necessity and possibility. Science becomes the art of the possible because the interesting questions are no longer how the world is, but how it could be, and how we can most effectively create other universes—given this or that set of computational resources.

Deconstructing the Subject Matter of Biology

Viewing research in artificial life as a postmodern science, where the computer is to the researcher what nature was to the classical natural scientist, leads us to render a reverse label to naive realism. We obtain a mirror image whereby the theory—or theories—of simulated life are a construction that is just as random as a virtual reality in a science-fiction–like cyberspace. It would make any kind of knowledge (any proposition about these virtualities) possible and equally valid, and this is in an elementary sense the same as antiscience.

Rather, we should consider the challenge of artificial life and complex systems as a means of breaking through several assumptions in established sciences from within, because this research forces us to rethink the entire project of traditional biology. Biology has been characterized by several partly metaphysical oppositions: between form and function, part and whole, heredity and environment, organism and environment, historical randomness and environmental necessity, energy and information, reduction and synthesis, mechanics and vitalism, concept and metaphor. These oppositions apply both to methods and to the complex objects themselves. The construction of artificial life can in a general way contribute to dissolving some of these oppositions, or combining them in new ways so as to shed new light on what living organisms really are; we may even use a-life to criticize the very opposition between life and death, between organic and inorganic—the essential opposition that has formed the basis for biology as a project. By creating organic simulacra and by their instability in relation to traditional concepts of concrete biological systems, artificial life may help contribute to a deeper understanding of the logic of life, and to a critical reevaluation of the logic of research, to which both experimental and theoretical biology are subordinated. Artificial life thus becomes a path toward what we might call a deconstruction of biological science, its scope, and subject matter.[115]

What must be deconstructed first is the naive view of biology (and of natural sciences generally) as unmediated depiction. Theories are never pure copies but always help to constitute the "objects" that they connect into a rational whole.[116] Theories utilize models of every kind as tools, and these models are projected onto the object's domain in order to compare the model's behavior with that of "reality." This applies regardless of whether we are speaking of purely verbal, informal models, physical models, or formal models that can eventually be simulated.

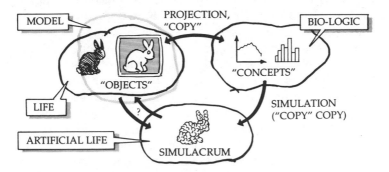

FIGURE 7.2. Biology as model life. Artificial life as the deconstruction of biology's traditional object.

Neither the theory nor the model can be understood as isolated simulacra. We might instead say that the scientific games of language create simulacra of the first order—that is, theories understood as a group of descriptions of certain previously selected and theoretically colored fragments of the world. A seemingly innocent concept like "an ecosystem" is an expression of a bio-logic projected onto a given phenomenon; every ecosystem in an ecological theory consists of equally abstract concepts like "populations," "the environment" and "biomass." The various fragments of the world are presented in a technical artificial language. What is special about the research program for artificial life and other computationally based sciences is the creation of second-order simulacra: that is, copies of the copies themselves. The instability (designated by the question marks in fig. 7.2) in our view of these models is created because they refer to a more abstract mode of viewing the object than do normal simulated models of given physical/biological systems. They have (more explicitly than we are used to) assumed the task of constituting the objects they imitate. The construction and the copy are here related to each other in yet another way, as the old problems of constructing models for the creation of form, evolution, or behavior are recast into a new biotheoretical space.

We cannot totally dispense with a reference to biological reality in one form or another. However, a-life simultaneously encourages us to ask fundamental questions about the very essence of biology. We have seen (in chapter 6) that in its purely computational form, the idea of a-life contains its own kinds of problems, associated with an unclear concept of computation. Nonetheless, artificial life can be seen as a daring proposal for a research project that is transcendental in the sense that it investigates the conditions for realization of information processing and biological processes generally in systems of every kind, while it is also empirical in the sense that it implements these conditions in models that can be tested for several operational properties.

In its constructive form, artificial life promises a dismantling of the struggle between holism and reductionism. At the same time, it contains tendencies toward both, either by reducing a living whole to a question of purely external behavior, or by making the technical synthesis of complex wholes the sole criterion for understanding. A degree of border crossing is necessary, crossing between a flat empirical and a dark, overly theoretical biology. New pathways must be found that can connect the microscope and the computer. In such a project, a degree of disciplinary promiscuity is necessary. We need periods where one discipline attacks the other; we need exchange and even theft of concepts, methods, and perspectives. And to continue our sexual metaphor, we need a dose of interdisciplinary unfaithfulness as well, perhaps some professional mate swapping, some kinky fantasies, and some healthy self-criticism that does not degenerate into sadomasochism. At the same time, we ought to be skeptical of any nonreflective combination of various interdisciplinary traditions. A deconstruction of biology must respect the multiplicity and the irreducible differences in science and in the world. It is a strategy that at once promotes a plurality of voices, an expansion of possible approaches to life, while at the same time maintains that it is impossible for us to put traditional biology and its arsenal of methods behind us.

The deconstruction of biology's object implies a renewed suspicion towards the very question of a homogeneous essence, a unified object, or an essence-of-life that characterizes life or living processes as such. It was the idea of the vitalists that the study of mechanics attempted to exorcize, in which organicist biology (of Waddington, Needham, Woodger, and others), represented a kind of compromise. Today, complex-system research is itself at risk of being driven into an advanced form of essentialist thinking with its continual assertions that life is a collective property in complex self-organized systems that can emerge from many media. Research in artificial life and simulated self-organization will perhaps contribute to disentangling some of these assumptions. To a certain extent, biology, via its traditional multiplicity of disciplines, has always contained arguments against an all too metaphysical view of its problem areas and empirical domains. The deconstruction of biology began with biology itself.

Artificial life can thus be seen as a contribution to the construction of a new biology, a biology whose contours remain hazy, and where labels such as "the biology of the possible," "synthetic biology," or "computational biology" may eventually reveal themselves to be misnomers. It is at once based on experience with the life we know and with experience that we do not have. It is both empirical and logical, computationally based research in those conditions that must be fulfilled in order to make possible the realization of living structures. Hopefully, this research will provoke considerations about the very conditions for doing biology. Perhaps life is a grand concept in a grand narrative, whose final chapter we do not know. Before we reach this final chapter, we have enough to deal with in taking care of the life we know.

·   ·   ·

We have seen the living game in its biological and artificial varieties. Life itself, of course, is something completely different. Artificial life will hardly mean anything for the way we experience a nightingale. The Emperor's playmaster in Hans Christian Andersen's tale was not mistaken when he remarked, "With a real nightingale one can never calculate what is coming, but in this artificial bird everything is settled. That is the way it is and no other! One can explain it; one can open it and make people understand where the waltzes come from, how they go, and how one follows up another."[117] And as is known, the playmaster wrote a work of five and twenty volumes about the artificial bird, volumes that were very learned and very long, full of the most difficult Chinese words, which people had to say that they understood, for otherwise they would be punched in the stomach.

NOTES

Preface

1. The address "Light and Life" from 1932 by Niels Bohr is printed in *Nature* 131 (1933): 421 and in Niels Bohr 1961: *Atomic Physics and Human Knowledge*, New York: Science Editions.

Chapter One

2. On computer viruses, see Fred Cohen 1987: "Computer viruses: Theory and experiments," *Computers and Security* 6:22–35; and Philip Elmer-DeWitt 1992: "Invasion of the data snatchers," *Time*, September 26, 1992. On Thomas Ray's organisms, see T. S. Ray 1992: "An approach to the synthesis of life," pp. 371–408 in *Artificial Life II*, ed. C. G. Langton et al., Redwood City, Calif.: Addison-Wesley.

3. The game "Life" is described in detail in Elwyn R. Berlekamp, John H. Conway, and Richard K. Guy 1982: *Winning Ways for Your Mathematical Plays*, vol. 2, pp. 817–50, London: Academic Press. "Life" was first described by Martin Gardner in his popular column "Mathematical Games" in *Scientific American* 223 (Oct. 1970): 120–23, and later in 224 (Feb. 1971): 112–17. A major study of the game and its philosophical implications is found in William Poundstone 1985: *The recursive universe: Cosmic complexity and the limits of scientific knowledge*, William Morrow, New York (paperback, Oxford University Press, New York, 1987). The philosopher Daniel C. Dennett uses the Life game in a provocative discussion of the epistemological status of patterns: "Real patterns," *The Journal of Philosophy*, 88 (1991): 27–51.

4. See John F. Cornell 1986: "Newton of the Grassblade? Darwin and the Problem of Organic Teleology," *Isis* 77:405–21.

5. The very existence of organisms reveals a specific way to realize a physical system—that is, as a mode that includes a series of aspects: biochemical-physiological; genetic; ecological; plus the organism's own integrative holistic aspect. This organism-based mode of realizing a physical system cannot be adequately described in the vocabulary of physics. It is ontologically equivalent

to the generation of a certain level for the biological laws of matter. The idea that reality is organized in levels, each with its own characteristic entities and properties, is found in many variants in science, philosophy, and systems theory. (See James K. Feibleman 1954: "Theory of integrative levels," *British Journal of the Philosophy of Science* 5:59–66; and C. Emmeche, S. Køppe, and F. Stjernfelt 1994: "Emergence and the ontology of levels," unpubl. ms.) In biology, this idea is related to the widely accepted organicist view, which neither believes that biology can be reduced to physics (as in classical mechanism), nor ascribes to living systems any "hidden" mysterious qualities (as in vitalism). In twentieth-century biology, organicism has had many adherents, including J. H. Woodger, J. Needham, P. Weiss, C. H. Waddington, E. Mayr, R. Lewontin, and R. Levins. See D. J. Haraway 1976: *Crystals, fabrics, and fields: Metaphors of organicism in twentieth-century developmental biology*, New Haven, London: Yale University Press.

6. On information and the origin of life, see F. Dyson 1985: *Origins of life*, Cambridge: Cambridge University Press; and B.-O. Küppers 1990: *Information and the origin of life*, Cambridge, Mass.: M.I.T. Press. On life as a language-like phenomenon, or life as a specific kind of information or sign system, see C. Emmeche 1991: "A semiotical reflection on biology, living signs and artificial life," *Biology and Philosophy* 6:325–40; C. Emmeche and Jesper Hoffmeyer 1991: "From language to nature: The semiotic metaphor in biology," *Semiotica* 84: 1–42; and C. Emmeche 1990: *Det biologiske informationsbegreb*, Aarhus: Forlaget Kimære.

7. The papers from the first international meeting on artificial life have been published in Christopher G. Langton, ed., 1989: *Artificial life: The proceedings of an interdisciplinary workshop on the synthesis and simulation of living systems held September 1987 in Los Alamos, New Mexico* (also Santa Fe Institute Studies in the Sciences of Complexity, Proceedings, vol. 6), Redwood City, Calif.: Addison-Wesley. This volume, and the following, should be the bible for those seeking a comprehensive view of the field on a technically sophisticated level. Papers from the second international workshop in February 1990 in Santa Fe have been published by Christopher G. Langton, Charles Taylor, J. Doyne Farmer, and Steen Rasmussen, ed., 1992: *Artificial life II* (also Santa Fe Institute Studies in the Sciences of Complexity, Proceedings, vol. 10), Redwood City, Calif.: Addison-Wesley. From the First European Conference on Artificial Life (ECAL91), held in Paris in December 1991, is published Francisco J. Varela and Paul Bourgine, ed., *To-*

ward a practice of autonomous systems, Cambridge, Mass.: M.I.T. Press. A very nice introduction to the a-life movement and its history is given in Steven Levy 1992: *Artificial life: The quest for a new creation*, New York: Pantheon.

8. One could call it "the strong claim of a-life," like the philosopher John Searle talking about (and criticizing) "strong AI"—that is, the version of artificial-intelligence research that not only seeks to model mental processes and make computer simulations of the brain, but that also claims that the brain *is* the hardware of the mind, and that a digital artificial-intelligence computer program, sufficiently complex, realizes intelligent, intrinsically meaningful processes. (See John R. Searle 1980: "Minds, brains, and programs," *Behavioral and Brain Sciences* 3:417–58, cf. C. Emmeche 1991: "The Problem of medium-independence in artificial life," pp. 247–57 in *Complexity, chaos, and biological evolution*, ed. Erik Mosekilde and Lis Mosekilde (NATO ASI Series B, vol. 270), New York: Plenum.) One can certainly be content with a more modest or "weak" version of the a-life program that does not embrace such claims as those listed under "real (artificial) life" "all life is form" (p. 18).

9. *Emergence*: The fact that properties of a system may be different from (and not easily explained by) the properties of the component subsystems. Emergence may also refer to the process of the generation of a phenomenon at a higher level, where the phenomenon's behavior cannot be deduced from knowledge of the lower level, even though it originates from complex interactions among the component parts (sometimes called collective behavior). The word emergence has been used in various senses since C. Lloyd Morgan published *Emergent Evolution* (London: Williams & Northgate) 1927 and Stephen Pepper wrote the paper "Emergence" in *Journal of Philosophy* 23 (1926): 241–45. See also R. L. Klee 1984: "Micro-determinism and concepts of emergence," *Philosophy of Science* 51:44–63; the papers by S. Salthe, G. Kampis, J. Fernandez, A. Moreno, A. Etxeberria, F. Heylighen, and others in *World Futures* 32 (1991); and David Blitz 1992: *Emergent evolution*, Dordrecht: Kluwer.

10. One normally has to use mathematical tools to describe precisely the whole aspects of a system's evolution, e.g., in nonlinear dynamics, which partly corresponds to the popular notion of chaos theory. A system's trajectory through a high-dimensional "phase-space" (each dimension characterizing some parameter of the system) can be described as belonging to one of several quali-

tatively different types of dynamics, some of which may show an extreme sensitivity to the initial conditions (which is the crucial condition for chaotic behavior in the mathematical sense). An entertaining introduction to chaotic dynamical systems is James Gleick 1987: *Chaos: Making a new science*, New York: Viking.

11. That it is not possible to predict the P-type from the G-type is here equivalent to saying that "no shortcut is possible" (S. Wolfram 1985: "Undecidability and intractability in theoretical physics," *Physical Review Letters* 54:735–38). What this means is that very often one cannot find a short formula F that gives the solution for a problem (corresponding to the system's final state, the P-type) with fewer computational steps than the number of steps used by the system itself in its step-by-step "computation" of its own evolution. This is related to the undecidability of the so-called Turing Halting Problem: Alan Turing showed that there exists no universal algorithmic procedure F (a general program) for deciding whether a "computer" M (a Turing machine), which is given an arbitrary input program G, will compute a definitive output P as a solution and then stop, or whether M will go on crunching numbers forever, never reaching a conclusion. As a rule, one simply has to wait and see (and one may wait forever!). See also S. Wolfram 1984c: "Universality and complexity in cellular automata," *Physica D* 10:1–35.

12. J. Doyne Farmer and Alletta d'A. Belin 1992: "Artificial life: The coming evolution," pp. 815–33 in Langton et al., ed., 1992 (see n. 7 above).

13. The Danish cognitive scientist Michael May (in "Kognitions-og kunstig intelligensforskning mellem logik og psykologi," *Psyke & Logos* 9 (1988): 253–74) has suggested distinguishing between AI as a psychological theory and AI as part of a general logical-semiotic theory of the general structures of thinking and its conditions of possibilities.

CHAPTER TWO

14. Denis Diderot 1769: *d'Alembert's Dream* (Harmondsworth: Penguin Books, 1976, pp. 158–59); here cited from p. 81 in Ilya Prigogine and Isabelle Stengers 1984: *Order out of chaos: Man's new dialogue with nature*, Toronto: Bantam Books.

15. The notion of a species-specific subjective *Umwelt*, which for the individual organism is "the niche seen from within," originated with the Estonian-German biologist Jakob von Uexküll

(1864–1944) and has been further developed by his son, Thure von Uexküll. See T. v. Uexküll 1982: "Introduction: Meaning and science in Jakob von Uexküll's concept of biology," *Semiotica* 42:1–24; and 1992: "Introduction: The sign theory of Jakob von Uexküll," *Semiotica* 89:279–315.

16. See, however, Milan Zeleny's comment on p. 20 in Zeleny, ed., 1981: *Autopoiesis: A theory of living organization*, North Holland: Elsevier. The possibility of man-designed (heteropoietic) artificial life may render the autopoietic/heteropoietic distinction problematic. See H. R. Maturana and Francisco J. Varela 1980: *Autopoiesis and cognition*, Dordrecht: Reidel; Maturana and Varela 1987: *The tree of knowledge: The biological roots of human understanding*, Boston: Shambala. A critique of Maturana's subjectivism is John Mingers 1990: "The philosophical implications of Maturana's cognitive theories," *Systems Practice* 3:569–84.

17. Aristotle's *Physics* 254b (here cited from p. 231 in Johannes Witt-Hansen 1965: *Kompendium til forelæsninger over den antikke filosofis historie*, 3d ed., Copenhagen: Munksgaard). A new edition of Aristotle's biological works is *History of Animals*, vol. 1–3, Cambridge, Mass.: Loeb Classical Library, Harvard University Press, 1937–1991.

18. The Danish psychologist Niels Engelsted is inspired by Aristotle when he claims that psyche—living intentionality—is not a collective or emergent property of large information-processing systems, but a quality that characterizes the active, *teleological* relationship between a system ("a general subject") and its environment (the "object"). E.g., when an organism uses energy *in order to* increase its own access to energy. (See Engelsted's contribution in N. Engelsted, L. Hem, and J. Mammen, ed., 1989: *Essays in general psychology*, Aarhus: Aarhus University Press).

19. Pallas, Johann Blumenbach, A.-L. de Jussieu, F. de Vicq d'Azyr, Goethe, and J.-B. Lamarck also embraced this point of view. See also chap. 2 in François Jacob 1973: *The Logic of Life*, New York: Pantheon (repr. 1993, Princeton: Princeton University Press).

20. S. J. Gould 1990: *Wonderful life: The Burgess Shale and the nature of history*, New York: Norton.

21. On SETI, see M. D. Papagiannis 1985: "Recent progress and future plans on the search for extraterrestrial intelligence," *Nature* 318:135–40. On exobiology, see Carl Sagan, "Life," in *The new encyclopædia britannica*, 15th ed., vol. 10, London: William Benton, 1943–1973; and G. Feinberg and R. Shapiro 1980: *Life beyond the earth*, New York: William Morrow.

22. Sagan, "Life."

23. Langton, "Artificial life," in Langton, ed., 1989 (see n. 7 above).

24. Claude Bernard 1878: *Leçons sur le phénomène de la vie communs aux animaux et aux végétaux* (trans. *Lectures on the phenomenon of life common to animals and plants*, Springfield, Ill.: C. C. Thomas, 1974, p. 24ff). In addition, Bernard claimed that life was not distinguished from inorganic phenomena by its physiochemical properties, but that the distinguishing feature of the living organism was the "definite idea" directing its development. He added, however, that "in physiology, it is necessary to renounce the illusion of a definition of life. We can only characterize its phenomena" (p. 44).

25. Pasteur thought incorrectly that the asymmetric molecules (that rotate the light) only could be produced by "vital principles" specific to living matter, and that asymmetry related to a "cosmic principle," which he thought was reflected in living beings. Nils Roll-Hansen 1974: *Forskningens frihet og nødvendighet* [Research's freedom and necessity], Oslo: Gyldendal Norsk Forlag.

26. See E. Schrödinger 1944: *What is life?* Cambridge: Cambridge University Press; and J. D. Watson 1968, *The double helix*, New York: Atheneum.

27. The list is modified from Carl Sagan (see n. 22 above).

28. This research and some of its consequences for natural philosophy is beautifully discussed by Prigogine and Stengers 1984 (see above, n. 14). They can be criticized, however, for the fact that they do not clearly distinguish between biotic and nonbiotic (purely physical) forms of self-organization. They thereby create the illusion that the theory can explain the complexity of the level of organization that characterizes, for instance, a bacterial cell. It is uncertain and not very likely that thermodynamic processes in open systems without any replication (i.e., systems with no intrinsic symbolic representation of the system itself, in the form of genetic information for instance) can lead to the complexity one observes in living beings.

29. In a certain sense this is a sixth definition of life, though it has the same problem as the thermodynamic one: it comes to include many purely physical systems. From a physical point of view, this is quite unproblematic: physics is certainly the science concerned with all that can be completely described in physical terms. However, the existence of other sciences (biology, psychology, sociology, etc.) indicates the existence of other forms of or-

ganization (compare n. 5 above) that have properties so specific and complex that they cannot be described physically. (On self-organized criticality, see n. 96 below.)

30. Vague should here be understood in C. S. Peirce's sense—that is, realistic. If there is a real continuum between the inorganic and the living, logic should not eliminate borderline cases. See Claudine Engel-Tiercelin 1991: *Peirce's logic of vagueness*, IMFUFA-text no. 205, Roskilde: Roskilde University; and 1990: *Le probléme des universaux*, doctoral thesis, University of Paris.

31. T. S. Ray 1992: "An approach to the synthesis of life," pp. 371–408 in Langton et al., ed., 1992 (see n. 7 above).

32. Emmeche and Hoffmeyer 1991 (see n. 6 above).

33. Lautrup's definition is not concerned with the ontological status of artificial life (the distinction between genuine artificial life and simulated artificial life, as in the third of the seven commandments), but with the level of the computer within which it unfolds itself, and with the consequences this has for the ability to control a-life. Thus, here "real/virtual" will designate Lautrup's distinction. One may see computer viruses as real artificial life (in Lautrup's sense), which only simulate real viruses, or one may see virtual organisms as instances of genuine life (as is probably the case for adherents of the strong version of the research program, see n. 8 above).

34. Ray's interpretation regarding metabolism presupposes that these viruses will not be defined purely as digital (so their specification in a purely symbolic code should be sufficient). This, however, seems to contradict Ray's own definition of his organism and the usual way of looking at computer viruses.

35. From a sign-theoretic, i.e., semiotic, point of view, life is a dynamic state of matter in which the code-duality between the analog and the digital is a crucial property, as is the ability to react selectively to signs from the environment. See Jesper Hoffmeyer and C. Emmeche 1991: "Code-duality and the semiotics of nature," pp. 117–66 in Myrdene Anderson and Floyd Merrell, ed., *On semiotic modeling*, Berlin: Mouton de Gruyter.

CHAPTER THREE

36. Or, from the point of view of modern industrialized agriculture we could ask: Do hens need wings at all in order to produce eggs? Even if the organism is not a man-made machine, the existence of cooped-up hens and caged calves supports the reduction-

istic metaphor: the organism is reduced to a cog in the agro-industrial machine!

37. "Organism" is derived from the same word as organ: in Latin, *organum*; in Greek, *organon*, which means tool, and was the title given to Aristotle's logical writings to emphasize the idea of logic as a tool helping the other sciences. An organism is, in the dictionary's sense, a system of organs (a system of tools). The French word *organisation* is another derivation. "Machine" is derived via the Latin *machina* from the Greek *mechane*, meaning tool, or machine, though not an innocent tool. The principal meaning of *mechane* is as an instrument (often in a negative sense) to lift heavy objects, a crane, or a military engine, something that could be turned against oneself. Perhaps the original word is *mëxos*, which means an artificial device, especially used against misfortune and troubles, a magical defense (known from Aischylos, about 500 B.C.). A century later it received the meaning "a mechanical device for the procreation of misfortunes," a contrivance to plot schemes; cf. *mëchanikos*, inventive, clever, especially at things evil. In this way the present difference in meaning between organism and mechanism—and the difference between nature's ingenuity and man's fabrication—has a long history. (I thank Laurits Lauritsen for this information.)

38. See Diderot 1769 (see n. 14 above). Diderot, a mechanical materialist and Enlightenment philosopher, was painfully aware of the limits of the mechanistic philosophy with respect to living beings.

39. See especially "Fifth lecture" from December 1949, printed in John von Neumann 1966: *Theory of self-reproducing automata*, Arthur W. Burks, ed., Urbana: University of Illinois Press; this edition also contains the large, uncompleted manuscript on the twenty-nine-state CA model and Burks's detailed commentary. Further introduction and discussion can be found in A. W. Burks, ed., 1970: *Essays on cellular automata*, Urbana: University of Illinois Press.

40. Even though Avery, MacLeod, and McCarty had in 1944 published results showing that DNA was the "transforming factor" that could change the strain-specific surface coat of pneumococcus bacteria, relatively few scientists drew the conclusion (which did not in all respects follow logically from their experiments) that genes were made only of DNA until Watson and Crick revealed their famous double-helix model in 1953.

41. More details are given in A. Chapuis and E. Droz 1958: *Automata: A historical and technological study*, trans. A. Reid, London: B. T. Batsford.

42. For a soda machine, a simple program could be "If input = $1, open box S, give in return $0, and rotate the wheel one unit; if input is less, return all input money; if input = $1 + *x*, open box S and give $*x* in return." Bayard Rankin and R. J. Nelson provide an excellent introduction to "Automata theory" in *Encyclopædia britannica*.

43. The real strength of modern computers is their ability to simulate another machine (writing, calculating, painting machines, etc.). Put another way, a computer is a *second-order machine* that—when given a formal specification of a *first-order machine* (e.g., a word processor)—itself becomes (simulates/realizes) this first-order machine.

44. The analogy was made by the physicist Freeman Dyson in 1970 (cf. p. 302 in A. W. Burks 1975: "Logic, biology and automata—some historical reflections," *Int. J. Man Machine Studies* 7: 297–312). It should be noticed that the analogy is not quite simple, because the "self-description" in the cell's DNA is not formal or logical (as with von Neumann), but is itself an integral part of the metabolism. Furthermore, component A constructs by following a specific sequence of operations, formalized in a symbolic code; in living organisms, what is explicitly coded in the genes is only the sequence of amino acids of the individual proteins. The very "construction" of the cell or organism depends on processes with a considerable self-organizing character. Thus, one can critically ask to what extent it really is "biological" self-reproduction von Neumann has formalized, or if it is only an abstract, wider concept of self-reproduction. See also the discussion in chap. 6 and n. 63; and G. Kampis and V. Csányi 1991: "Life, self-reproduction and information: Beyond the machine metaphor," *J. theor. Biol.* 148:17–32; and C. Emmeche 1992: "Modeling life: A note on the semiotics of emergence and computation in artificial and natural living systems," pp. 77–99 in Thomas A. Sebeok and Jean Umiker-Sebeok, ed., *Biosemiotics: The semiotic web 1991*, Berlin: Mouton de Gruyter.

45. This simulation does not therefore become self-reproducing: that would be a model of von Neumann's model. In the first essay in the volume *Essays on cellular automata* (see n. 39 above) Burks provides a survey of the construction of the various "organs" in the twenty-nine-state automaton; he explains how finite automata

and Turing machines are embedded in this construction; and finally he sketches universal construction and—as a special case of that—self-reproduction.

46. Christopher G. Langton 1984: "Self-reproduction in cellular automata," *Physica D* 10:135–44.

47. Langton, "Artificial life," p. 11 in Langton, ed., 1989 (see n. 7 above).

48. von Neumann 1966 (see n. 39 above), p. 77.

49. See Langton 1984 (n. 46 above).

50. Martin Gardner, "The fantastic combinations of John Conway's new solitaire game "life," *Scientific American* 223 (Oct. 1970):120–23 (see n. 3 above).

51. R. W. Gosper, here cited from "Signs of life: Transcript of the programme transmitted 11th June, 1990," produced by John Wyver, Horizon, London: BBC.

52. *Scientific American* 224 (Feb. 1971): 112–17.

53. An algorithm to explore problems of the same type as the "puffer train," see R. W. Gosper 1984: "Exploiting regularities in large cellular spaces," *Physica D* 10:75–80.

54. A. K. Dewdney, *Scientific American* 256 (Feb. 1987): 8–13.

55. See n. 51 above.

56. Poundstone 1985, chap. 12 (see n. 3 above).

CHAPTER FOUR

57. The idea of mathematics as a kind of experimental science is well described in David Campbell, Jim Crutchfield, J. Doyne Farmer, and Erica Jen 1985: "Experimental mathematics: The role of computation in nonlinear science," *Communications of the ACM* [Association for Computing Machinery] 28:374–84.

58. D'Arcy Wentworth Thompson 1917: *On growth and form*, Cambridge: Cambridge University Press (2d. ed., New York: Macmillan, 1942); see also S. J. Gould 1976: "D'Arcy Thompson and the science of form," *Boston Studies in the Philosophy of Science* 27:66–97.

59. For a more detailed and formally correct description of L-systems, see A. Lindenmayer and P. Prusinkiewicz, "Developmental models of multicellular organisms: A computer graphics perspective" in Langton, ed., 1989 (see n. 7 above); and P. Prusinkiewicz and A. Lindenmayer 1990: *The algorithmic beauty of plants*, New York: Springer Verlag. For the enthusiast who wants to experiment with self-similar curves on a desktop computer, more inspiration can be found in B. B. Mandelbrot 1982: *The fractal geome-*

try of nature, San Francisco: Freeman. The Barnsley fern in fig. 4.3 is described in Gleick 1987 (see n. 10 above); see also Barnsley 1985: "Iterated function systems and the global construction of fractals," *Proc. Royal Soc. Lond. A* 339:243–75.

60. Programming by using recursive rules, i.e., rules that can be applied several times to an object, such as a mathematical expression, and that each time change the object accordingly. Recursiveness is also used in logic, where, for instance, "If F is a well-formed formula, then −F is also a well-formed formula" is a recursive rule; or in mathematics, where "$n! = n*(n-1)!$; $0! = 1$" is a recursive definition of the factorial function: one calculates $n!$ by multiplying n with $(n-1)!$, and the latter expression is calculated by applying the same rule, $(n-1)! = (n-1)*(n-2)!$, and so on, hence $5! = 5*4! = 5*4*3! = 5*4*3*2! = 5*4*3*2*1 = 120$.

61. The computer scientist C. H. Bennett has suggested a measure for a structure's degree of *complexity*, namely its logical depth. Logical depth is defined as the time needed (measured as number of computational steps) for the shortest possible program to generate the structure; i.e., the time consumption from the input (the minimal algorithm) to the resulting output (Bennett 1986: "On the nature and origin of complexity in discrete, homogeneous, locally interacting systems," *Foundations of Physics* 16:585–92). A true deep structure is thus characterized by the mathematical property that it cannot be generated faster (by fewer computational steps) via a simulation on any other computer.

62. P. Oppenheimer, "The artificial menagerie," pp. 251–74 in Langton, ed., 1989 (see n. 7 above).

63. The *genetic* code is the triplets, that is, "words" of three letters, constituted by the nucleotide bases A, G, C, and T located along the DNA backbone, each triplet coding for an amino acid (which is part of the specific protein that the gene in question codes for). There is an enormous jump in complexity from one gene's code for a single protein (that can be determined as a detailed DNA sequence) to the "code" for the structures formed by the cells of an organism. The latter has been termed the epigenetic code: it must contain a partial description of epigenesis (the embryonic development: growth, cell differentiation, and morphogenesis), which as a coherent and complex process is due to self-organizing as well as cell-biological processes. Their specific relation to the intracellular coding of individual proteins is far from fully known in all details. Even the self-assembly of the individual protein is not explicitly described in DNA (as is the pri-

mary sequence of its amino acids), and therefore, the notion of plant development as algorithmic should be seen as a metaphor with limited validity. This form-representation problem in biology is similar to the problem of knowledge representation in cognitive science. On biological information economy, see R. Riedl 1977: "A systems-analytical approach to macro-evolutionary phenomena," *The Quarterly Review of Biology* 52:351–70. On epigenetic and pre-formationistic metaphors for generation of form, see S. Oyama 1985: *The ontogeny of information*, Cambridge: Cambridge University Press.

64. See note 63.

65. The expression "coded in the whole" is a metaphor for a system (P) with dynamic emergence of macroproperties, produced from local interactions between elements on the microlevel. This need not presuppose that we are dealing with a system (B), constituted by a distinction between an object, its description, and a code giving the rules for writing (generation of the description) and reading (interpretation of the description). Many self-organizing systems of type P contain no form of coded self-description, while a living cell belongs to both the class P of dynamic systems with emergence and to class B of systems that contain their own description.

66. Though there are versions of the Life game where living cells take on an increasingly darker shade of red as the number of generations they have been turned on increases.

67. On use of cellular automata for simulating physical processes, see Tommaso Toffoli and Norman Margulis 1987: *Cellular automata machines: A new environment for modelling*, Cambridge, Mass.: M.I.T. Press; and A. K. Dewdney's discussion in *Scientific American* 262 (Aug. 1989):88–91.

68. The geometry metaphor is from C. H. Waddington 1968: "The basic ideas of biology," pp. 1–32 in Waddington, ed.: *Towards a theoretical biology*, vol. 1: *Prolegonema*, Edinburgh: Edinburgh University Press. The epigenetic landscape is from Waddington 1957: *The strategy of the genes*, London: George Allen & Unwin. See also Scott F. Gilbert 1991: "Epigenetic landscaping: Waddington's use of cell fate bifurcation diagrams," *Biology & Philosophy* 6:135–54.

69. There are also elements of *disanalogy* in this analogy between (automaton rules + initial state) and (DNA program + cellular structures [epigenetic substrate]). First, the separation between programmed rules and data is not so sharp in the biological cell where the epigenetic substrate both resembles the input data upon

which DNA must operate and the very machine that is supposed to read and interpret DNA: the hierarchy of the cell is tangled. Second, the analogy is not so compelling because one might allow the automaton rules to correspond to (a) the biological structures that "run" DNA; and (b) the initial state of the automaton to the biological genotype.

70. The qualitative classes are analogous to the types of behavior in dynamic systems, which develop towards, respectively, (1) a point attractor, (2) a limit cycle, (3) a chaotic attractor, (4) an attractor with very long transients (correlations over time). A comprehensive study of such attractors is A. Wuensche and M. J. Lesser 1992: *The global dynamics of cellular automata: An atlas of basin of attraction fields in one-dimensional cellular automata*, Santa Fe Institute Studies in the Sciences of Complexity, Reference vol. I, Redwood, Calif.: Addison-Wesley.

71. S. Wolfram 1984a: "Computer software in science and mathematics," *Scientific American* 251 (Sept.): 140–51; and 1984b: "Cellular automata as models of complexity," *Nature* 311:419–24; and 1984c, "Universality and complexity," *Physica D* 10:1–35. One should be aware of the fact that cellular automata type simulations impose space-time granularity and synchronicity that seriously may compromise their ability to describe the real world, see Bernardo A. Huberman and Natalie S. Glance 1993: "Evolutionary games and computer simulations," *Proc. Natl. Acad. Sci. USA*, 90: 7716–18; see also William Brown 1993: "Cheat thy neighbor—a recipe for success," *New Scientist* 140 (no. 1894): 19.

72. See, for instance, Georg M. Malacinski and Susan V. Bryant, ed., 1984: *Pattern formation*, New York: Macmillan.

73. The model is described in Craig W. Reynolds 1987: "Flocks, herds, and schools: A distributed behavioral model" (Proceedings of SIGGRAPH 1987), *Computer Graphics* 21:25–34. Langton's comment is from his paper "Artificial Life" in Langton, ed., 1989 (see n. 7 above). For a recent study of real birds' flocking behavior from this perspective, see G. De Schutter and E. Neuts 1993: "Birds use self-organized social behaviours to regulate their dispersal over wide areas: Evidences from gull roosts," in Proceedings from the second European conference on artificial life (ECAL93, in Brussels, Belgium).

74. It is different for plants because they are probably more capable of surviving, for instance, doublings of the chromosome number of the cell from one generation to the next, literally creating a new species. See D. Briggs and S. M. Walters 1984: *Plant*

variation and evolution, 2d ed., Cambridge: Cambridge University Press.

75. The Sophist Protagoras of Abdera used the same argument with respect to the possibility of knowing the gods.

76. Neo-Darwinism as a paradigm embraces a view of evolution as a slow, gradual process, a view criticized by S. J. Gould, N. Eldredge (see their "Punctuated equilibria comes of age," *Nature* 366 (1993): 223–27), and other advocates of punctuated equilibria. The latter constitutes an alternative pattern of evolution where long stable periods of stasis are interrupted by periods of rapid speciation. For an introduction to this discussion see G. L. Stebbins and F. J. Ayala 1985: "The evolution of Darwinism," *Scientific American* 253 (July 1985): 54–64; J. Maynard Smith 1988: "Punctuation in perspective," *Nature* 332:311–12; and the special issue of *Journal of Social and Biological Structures* 12 (1989):105–302, on punctuated equilibria.

77. K. Lindgren 1990: "Evolution in a population of mutating strategies," Nordita Preprint 90/22 S (Nordita, Copenhagen); and K. Lindgren 1992: "Evolutionary phenomena in simple dynamics," pp. 295–312 in Langton et al., ed., 1992 (see n. 7 above). Strictly speaking, Lindgren's model is not macroevolutionary because it describes (micro)evolution within a population of individuals consisting of different genotypes, not really a species in the biological sense. Furthermore, it contains a set of simplifying assumptions, which means that it cannot be interpreted as a very realistic model of microevolution within a population. However, the great value of the model is that it facilitates understanding of general evolutionary phenomena as stasis and punctuated equilibria, that is, phenomena that are naturally occurring within the abstract space of complex dynamic systems. At the Second European Conference on Artificial Life (ECAL93), Lindgren presented an extended, two-dimensional version of his and Nordahl's model.

78. The dilemma concerns individual versus mutual total benefits, and arises in situations of which the following is a sort of paradigm: you and your accomplice (whom you don't care especially much for) have committed a crime. You have both been thrown in jail, in two separate cells, and are unable to communicate. You fearfully await the trial. The prosecutor offers each of you the following deal (and informs you both that the same deal is offered to both of you): "We have substantial evidence on you both. If both of you claim innocence ("cooperate"), we'll convict you anyway,

and you'll both get two years in jail (3 points). If you help us and admit your guilt and make it easier for us to convict your accomplice ("defect"), we let you free (5 points) and put him in for five years (0 points). But if you both claim guilty, we will give you both four-year sentences (1 point)." So the structure of the situation is given by a scheme, whereby a pair such as (5, 0) denotes that player 1 gets 5 and player 2 gets 0.

Player 2

		Cooperate	Defect
Player 1	Cooperate	(3, 3)	(0, 5)
	Defect	(5, 0)	(1, 1)

You do not know what your partner chooses, so what will you choose? (Logic says that for a single game, it is rational for both players to defect, even though it would be to the players' mutual advantage to establish cooperation in the long run.) The problem, formulated in 1950 by Melvin Dresher and Merrill Floyd, is now a classic in game theory. Robert Axelrod and W. D. Hamilton have analyzed its consequences for evolutionary theory in *Science* 211 (1981): 1390–1996; see also Axelrod 1984: *The evolution of cooperation*, New York: Basic Books; and chap. 29 of D. R. Hofstadter 1985: *Metamagical Themas*, New York: Basic Books.

79. It has been difficult to prove that an ecological system, which is apparently in balance, should by itself be able to change dramatically. Therefore catastrophes (large volcanic eruptions or falling meteors) have been so attractive as hypotheses for explaining mass extinction of species at specific geological epochs, such as at the Cretaceous/Tertiary boundary about 65–70 million years ago.

80. See n. 77 above.

81. Steen Rasmussen 1990: "Er computervirus levende?" [Are computer viruses alive?] *Omverden* 2, Copenhagen; Steen Rasmussen, Carsten Knudsen, Rasmus Feldberg, and Morten Hindshold 1990: "The coreworld: Emergence and evolution of cooperative structures in a computational chemistry," *Physica D* 42:111–34. The model and the instructions used are a modification of a popular computer game, Core War, described by A. K. Dewdney in *Scientific American* 250 (May 1984): 15–19; and in vol. 252 (March 1985): 14–19.

82. Ibid.

83. Personal communication.

84. Rasmussen 1990 (see n. 81 above).

85. See for instance Brian Goodwin and Peter Saunders, ed., 1989: *Theoretical biology: Epigenetic and evolutionary order from complex systems,* Edinburgh: Edinburgh University Press.

86. Introductions to Stuart Kauffman's work include Kauffman 1987: "Dueling selectively with Darwin," *The Scientist* (Aug. 10), p. 21; Kauffman 1987: "Developmental logic and its evolution," *BioEssays* 6: 82–87; and M. M. Waldrop 1990: "Spontaneous order, evolution, and life," *Science* 247:1543–45. A more detailed discussion is Kauffman 1985: "Self-organization, selective adaptation, and its limits," 169–207 in David J. Depew and Bruce H. Weber, ed., *Evolution at a crossroads: The new biology and the new philosophy of science,* Cambridge, Mass.: M.I.T. Press. See also Kauffman 1993: *The origins of order: Self-organisation and selection in evolution,* New York: Oxford University Press.

87. Compare the critique of neo-Darwinism (see n. 76 above). Together with Simon Levin, Kauffman has made a plausible case for the idea that natural selection can be restricted by two "complexity catastrophies" that fundamentally limit the optimizing power of natural selection (Kauffman and Levin 1987: "Toward a general theory of adaptive walks on rugged landscapes," *J. theor. Biol.* 128:11–45). The first limit arises because the power of natural selection to optimize towards high fitness (given a constant mutation rate) gets smaller as the number of genetic loci increases. Eventually mutation overwhelms selection and disperses an adapting population away from the optimal genotypes. (Classical population genetics has long hinted at this limit.) The other limit is due to a condition found by Levin and Kauffman, that when the model's genetic networks, due to selection, get more complex, the fitness peaks (of the rugged adaptive landscape reachable by the population) get lower. This may imply that even strong selection cannot by itself achieve highly complex and precise systems.

CHAPTER FIVE

88. Virtual (from Low Latin *virtualis,* from *virtus,* power, ability): (a) the original meaning seems to be something having the inherent power to produce certain effects; something that can act or become present though is not yet actually so; a potential capacity; (b) in common usage something real (as opposed to purely nominal), a matter of fact, though not expressed or appearing as such (e.g., a virtual defeat, not declared); (c) in the development

of physics the term obtained its technical meanings ("virtual displacement" in Jordanus de Nemore's statistical mechanics, or "virtual work"; "virtual picture" in optics) and in computer science we have seen the term applied to simulated automata. Thus, its meaning has been transformed from (a) something potential (opposed to actual); (b) something real, factual; to (c) something imagined, thought, or simulated that is realized in an artificial reality or hyperreality.

89. Jordan Pollack points to this as a problematic parallel between a-life and AI research: In a-life, too, there exists a variety of submodels, in which one describes only restricted parts or aspects of life (like intelligence in AI), e.g., self-reproduction, differentiation, evolution, morphogenesis. It is very difficult to attain a state of the art where agreement can be reached about calling a model really alive (see Richard K. Belew 1991: "Artificial life: A constructive lower bound for artificial intelligence," *IEEE Expert* 6:8–15, and "ALife-2: The second artificial life conference," ibid.: 53–59.

90. Ibid.

91. J. H. Holland 1975: *Adaptation in natural and artificial systems*, Ann Arbor: University of Michigan Press. Shorter introductions are given in A. K. Dewdney 1985: "Computer recreations: Exploring the field of genetic algorithms in a primordial computer sea full of flibs," *Scientific American* 253 (Nov.): 16–21; and C. T. Walbridge 1989: "Genetic algorithms: What computers can learn from Darwin," *Technology Review* (Jan.), pp. 47–53, or see Langton, ed., 1989 (see n. 7 above). A must for the beginner as well as the specialist is J. R. Koza 1992: *Genetic programming: On the programming of computers by means of natural selection*, Cambridge, Mass.: M.I.T. Press.

92. Richard K. Belew, John McInerney, and Nicol N. Schraudolph 1992: "Evolving networks: Using the genetic algorithm with connectionist learning," pp. 511–47 in Langton et al., ed., 1992 (see n. 7) above.

93. E.g., *New Scientist* (1982), Jan. 14., pp. 68–71; and Jan. 28., p. 236; J. B. Tucker 1984: "Biochips: Can molecules compute?" *High Technology* (Feb.), pp. 36–47; and R. C. Haddon and A. A. Lamola 1985: "The molecular electronic device and the biochip computer: Present status," *Proc. Natl. Acad. Sci. USA* 82:1874–78.

94. Yojiro Kondo 1989: "Introduction of bio-electronic device project," pp. 137–40 in 8th Symposium on Future Electronic Devices, Oct. 30–31, 1989, Tokyo.

95. D. R. Hofstadter and Paul Smolensky have provided argu-

ments for this view, cf. Smolensky's papers in *Artificial Intelligence Review* 1 (1987): 95–109; and his "On the proper treatment of connectionism," *Behavioral and Brain Sciences* 11 (1988): 1–74.

96. Langton's Ph.D. thesis, summarized in Langton 1992, "Life at the edge of chaos," pp. 41–91 in *Artificial Life II* (see n. 7 above), and in Langton 1990: "Computation at the edge of chaos: Phase transition and emergent computation," *Physica D* 42:12–37. This issue of *Physica D*, ed. Stephanie Forrest, is entitled *Emergent computation*, and contains proceedings from "The Ninth Annual CNLS Conference" in Los Alamos on this topic. Some physicists have recently (and apparently before Langton) found similar indications of the emergence of complex phenomena in the critical transition zone between order and disorder, a principle termed "self-organized criticality." See Per Bak et al. 1988: "Self-organized criticality," *Physical Review A* 38: 364–74; Per Bak et al. 1989: "Self-organized criticality in the 'game of life'" *Nature* 342:780–82; A. Mehta and G. Baker 1991: "Self-organising sand pile," *New Scientist*, 15 June, pp. 40–43.

97. Langton 1990 (see n. 96 above), p. 35. What Langton has in mind is probably the membrane of the cell and its organelles, in which phase-transition phenomena have been shown to exist. (See M. Bloom, I. Evans, and O. G. Mouritsen 1991: "Physical properties of the fluid bilayer component of cell membranes," *Quarterly Review of Biophysics* 24:293–397.) Such phenomena may also appear in the cytosole (that half part of the cytoplasm that does not consist of organelles). About twenty percent of the weight of the cytosole is proteins, so it is more a gelatinous mass than a simple solution of macromolecules.

98. Langton 1990 (see n. 96 above), p. 31, emphasis in original. Here, Langton is very close to postulating the emergence of a form of protosemantic information in this otherwise purely physical (or, strictly speaking, "logical-syntactical") system. It is an open question if such a postulate is false due to what Searle claims about the relationship between physics and computation; namely, that "syntax is not intrinsic to physics" (in Searle 1992: *The rediscovery of the mind*, Cambridge, Mass.: M.I.T. Press). Searle's argument that nothing in itself is a digital computer (because this requires ascriptions by an observer) appears convincing; however, one might conceive of a broader concept of analog computation where emergent symbols have an intrinsic (not just derived) syntax and even meaning (cf. C. Emmeche: "The idea of biological computation," in prep.).

99. On Peirce, see C. S. Peirce 1955: *Philosophical writings*, ed. Justus Buchler, New York: Dover. On biosemiotics, see Jesper Hoffmeyer, in press: "Semiotic aspects of biology: Biosemiotics," in Robert Posner, Klaus Robering, and Thomas A. Sebeok, ed.: *Semiotics: A handbook of the sign-theoretic foundations of nature and culture*, Berlin and New York: Mouton de Gruyter; and Thomas A. Sebeok and Jean Umiker-Sebeok, ed., 1992 (see n. 44 above).

100. Real computers (of today at least) are all irreversible, i.e., their computations require the discarding of information, a process that dissipates energy in the form of heat. As E. Fredkin, C. H. Bennett, and R. Landauer have pointed out, one could ideally (though probably not really) build fully reversible computers. If so, the energy consumption and heat dissipation per logical step could apparently be minimized to arbitrarily small amounts, which, however, presupposes that we can use an arbitrarily long time to do the computations. These considerations are, as one can see, theoretical, but nevertheless significant for our understanding of computation. The questions of the *absolute* physical limits on computation are still unresolved, but they are more likely related to cosmological questions, such as "Can we build computers that in the long run can circumvent the consequences of the Second Law of Thermodynamics?"; "How big a memory can we give a computer?"; "Will a logical operation require a minimal amount of time?"; "What are the tiniest possible devices that can do this kind of operations?" etc. See C. H. Bennett and R. Landauer 1985: "The fundamental physical limits of computation," *Scientific American* 253 (July): 48–56; Landauer 1988: "Dissipation and noise in computation and communication," *Nature* 335:779–84; W. Perod et al., 1984: "Dissipation in computation," *Physica Scripta* 52:232–35; H. M. Hastings and S. Waner 1985: "Low dissipation computing in biological systems," *BioSystems* 17:241–44; T. D. Schneider 1991: "Theory of molecular machines, I–II," *J. theor. Biol.* 148:83–137.

101. Landauer 1986: "Computation and physics: Wheler's meaning circuit?" *Foundations of Physics* 16:551–64; and Landauer 1987: "Computation: A fundamental physical view," *Physica Scripta* 35:88–95.

102. Feynman, cited by Robert Wright 1985: "The On-Off Universe," *The Sciences* 25 (Jan./Feb.): 7–9. Feynman's dictum recalls Peirce, who acknowledged that nature might not follow the physical laws with absolute precision, so as to stray a little bit away from the very laws or "habits" it has evolved itself.

103. On Fredkin, see Wright 1985 and R. Wright 1988: *Three scientists and their gods*, New York: Times Books; see also Julius Brown 1990: "Is the universe a computer?" *New Scientist* 14 July, pp. 37–39; Heinz R. Pagels 1988: *The dreams of reason: The computer and the rise of the sciences of complexity*, New York: Simon & Schuster; and E. Fredkin 1990: "Digital mechanics," *Physica D* 45:254–70.

Chapter Six

104. A critique of this tendency of the System-world to incessantly violate the borders between the System and the life world is succintly expressed by the Danish biochemist and philosopher Jesper Hoffmeyer. From his 1975 book, *Dansen om guldkornet*, Copenhagen: Gyldendal, p. 14, my translation: "The understanding of the living nature that the idea of objectivity has conferred on us seems to be more suitable to defend us against nature than to live with it."

105. T. Roszak 1972: *Where the wasteland ends: Politics and transcendence in postindustrial society*, London: Faber & Faber, p. 96.

106. I have not tried to distinguish the philosophical from the biological critique. The most important objections to artificial life (in the version that claims to achieve or realize genuine life computationally) have been presented by H. H. Pattee, "Simulations, above realizations, and theories of life," in Langton, ed., 1989 (see n. 7 above); Peter Cariani, "Emergence and artificial life," in Langton et al., ed., 1992 (see n. 7 above); R. Rosen, "What does it take to make an organism?" unpubl. ms.; Rosen 1991: *Life itself*, New York: Columbia University Press; Elliott Sober "Learning from functionalism: Prospects for strong AI," in Langton et al., ed., 1992 (see n. 7 above); and G. Kampis and V. Csányi 1987: "Replication in abstract and natural systems," *BioSystems* 20:143–52; see also Kampis and Csányi 1991 (see n. 44 above). Also, R. Penrose has made critical assertions on cellular automata as basis for biological models ("Signs of Life" transcript, see n. 51 above). The critique that Searle has put forward against strong AI (cf. the debate between Searle and P. M. Churchland and P. S. Churchland in *Scientific American* 262 (Jan. 1990): 19–31, may have analogous consequences for the question of medium independence of artificial life, even though one cannot make a complete parallel. See also C. Emmeche 1991 (see n. 8, above).

107. The key subjects in Mayr's list include complexity and organization; chemical uniqueness; quality; biological uniqueness

and variability; the possession of a genetic program written in DNA; historicity; natural selection; and indeterminism. See E. Mayr 1982: *The growth of biological thought*, Cambridge, Mass.: Harvard University Press, p. 53ff.

108. See, e.g., Geoff Simons 1985: *The biology of computer life*, Sussex: Harvester Press; and Langton, ed., 1989 (see n. 7 above). Cf. Langton's own paper and several others in this volume.

109. Even though Searle's Chinese Room argument against strong AI (see n. 106, above) may not be logically sound in a strict sense (though still intuitively convincing), and his biologism (or "biological naturalism"; see Searle 1992, see n. 98 above) with respect to mental states is debatable, I think that the argument against strong a-life can nevertheless be made over a similar, though not quite analogical, scheme.

110. Peter Cariani 1992 (see n. 106 above); Cariani 1989: "Adaptivity, emergence, and machine-environment dependencies," *Proc. of the 33rd Annual meeting of the Int'l Soc. for Systems Sciences* (formerly ISGSR), Edinburgh, Scotland, July 2–7, vol. 3, pp. 31–37; Cariani 1991: "Adaptivity and emergence in organisms and devices," *World Futures* 32:111–32. Cariani's ideas are inspired by the works of H. H. Pattee and R. Rosen.

111. This is true only to the extent that the training set and the performance measure, on which the computational changes depend, are external in relation to the observer's model. In a period without new training, the weights between the computing units of the network will be unchanged, and the network will not appear emergent to the observer in reference to his model.

112. Elliott Sober 1991, (see n. 106 above).

<h2 style="text-align:center">CHAPTER SEVEN</h2>

113. If one can at all talk about a diagnosis of society. Lyotard and his colleagues have been very careful to emphasize that their concepts should not be viewed as an attempt to characterize a specific phase or epoch that succeeds the modern epoch. Rather, they attempt to offer a radical critique of any form of quasiteleological philosophy of history, and to find an alternative mode of considering and analyzing the modern condition. An introduction to the discussion about postmodernism is given by Frederic Jameson 1990: *Postmodernism, or the cultural logic of late capitalism*, Durham, N.C.: Duke University Press.

114. These kind of questions are posed in relation to the so-

called anthropic principle; see John D. Barrow and Frank J. Tipler 1986: *The anthropic cosmological principle*, Oxford: Clarendon Press. A clear introduction to the discussion is given by Georg Gale in *Biology & Philosophy* 2 (1987): 475–91.

115. This deconstruction is analogous to that currently discussed in art, architecture, and literature. It tries to search for the various internally opposing voices or purposes in a text or an edifice, in a constructive way to allow traditional metaphysical oppositions to play against each other in new ways. See Christopher Norris 1991: *Deconstruction: Theory and practice*, London: Routledge.

116. The Danish philosopher Stig Andur Pedersen's example of a pear might be illustrative. A pear can be an object of a biochemical investigation. Here, one will abstract away all characteristics of the fruit which are not relevant to the investigation. For instance, one will ignore that the fruit is a commodity with a certain price on the fruit market. But the pear can also be an object of an economic investigation. In this case, one ignores all of its biochemical characteristics and looks only at the pear as a commodity. Objects of *any* scientific analysis are lifted out of their complex of natural elements and are idealized; they are made artifacts. It is the type of analysis that defines which aspects of the object are to be analyzed.

117. Hans Christian Andersen: "The Nightingale," p. 346 in *The Complete Illustrated Stories of Hans Christian Andersen*, London: Chancellor Press, 1983.

INDEX

Adam (evolutionary strategy), 97
adding machines, 52
agriculture, 173–74n.36
Aischylos, 174n.37
algorithms, 40, 78, 129. *See also* genetic algorithms
A-Life Electronic Mail Network, 148–49
allolactose, 104
amino acids, 33, 40–41, 175n.44, 177–78n.63
amoebae, 36, 45, 46
amphibians, 85
Andersen, Hans Christian, vii, 52, 166
animals: autonomy of, 23; domestic, 173–74n.36; environments of, 114; induction in, 112; macroevolution of, 91–92; reproduction of, 47; rights of, 32; simulated energy resources of, 40
animats, 112
ants, 12, 113–14
Aphrodite, 63
Aplysia, 131–32
Apollo 11 moon samples, 31
Apple Computers (firm), 16
"Approach to the Synthesis of Life, An" (Ray), 4
architecture, 159, 188n.115
Aristotle, 25–26, 27, 28; Engelsted and, 171n.18; form/matter complementarity and, 144, 149; on logic, 174n.37
art, 159, 188n.115
artificial chemistry, 99–102
artificial evolution, 6, 44, 91–98
artificial intelligence, 4, 13, 112–14, 138–39, 183n.89; human reasoning and, 21; information processing in, 19; Langton loop and, 148;

May on, 170n.13; robotics and, 111; Searle on, 169n.8; Turing test of, 142. *See also* neural networks
Artificial Life (Langton), 17
artificially modified life, viii–ix, 32–33, 41–42, 119
artificial physics, 123, 153, 160–61
asexual reproduction, 47, 48
assembler languages, 40, 41
astrophysical cosmology, 30
atavistic cell types, 108
atomic bomb development program, 51
attractors, 85–86, 105–6, 107, 108, 179n.70
automata, 51–60, 62, 153. *See also* cellular automata
automated translation, 53–54
automobiles, 34
autonomy, 23, 25, 45. *See also* movement
autopoiesis, 24–25, 34, 37, 141, 171n.16

bacteria, 34, 91, 174n.40. *See also* computer bacteria
Barnsley, M., 76
basins of attraction, 106
Baudrillard, Jean, 158
behavior: of artificial life forms, 18, 86; collective, 169n.9; cooperative, 94, 97, 101, 125; in dynamic systems, 98, 179n.70; information-determined, 113; programming of, 19
behavioral testing, 142
Belew, Richard K., 111, 117
Belin, Aletta d'A., 38, 39, 43–44, 46
Bennett, C. H., 177n.61, 185n.100
Bernard, Claude, 33, 172n.24